India as a Pioneer of Innovation

India as a Pioneer of Innovation

EDITED BY
HARBIR SINGH,
ANANTH PADMANABHAN,
AND
EZEKIEL J. EMANUEL

OXFORD
UNIVERSITY PRESS

OXFORD
UNIVERSITY PRESS

Oxford University Press is a department of the University of Oxford.
It furthers the University's objective of excellence in research, scholarship,
and education by publishing worldwide. Oxford is a registered trademark of
Oxford University Press in the UK and in certain other countries.

Published in India by
Oxford University Press
2/11 Ground Floor, Ansari Road, Daryaganj, New Delhi 110 002, India

First Edition published in 2017

ISBN-13: 978-0-19-947608-4
ISBN-10: 0-19-947608-X

Typeset in Dante MT Std 10.5/14.5
by The Graphics Solution, New Delhi 110 092
Printed in India by Replika Press Pvt. Ltd

Contents

Tables and Figures

Tables

Figures

Abbreviations

AIADMK	All India Anna Dravida Munnetra Kazhagam
AIDS	Acquired Immunodeficiency Syndrome
AIM	Atal Innovation Mission
AIOCD	All India Organisation of Chemists and Druggists
AIS	Agricultural Innovation Systems
ASHA	Accredited Social Health Activist
ATL	Atal Tinkering Laboratory
ATM	Automated Teller Machine
BAIF	Bhartiya Agro Industries Foundation
BBMP	Bruhat Bengaluru Mahanagara Palike
BG	Business Group
bn	billion
BPL	Below Poverty Line
CCU	Coronary Care Unit
CHW	Community Health Worker
COIN	Co-Innovation
CRISPR	Clustered Regularly Interspaced Palindromic Repeats
CT	Computerized Tomography
CTO	Chief Technology Officer
DFID	Department for International Development
DNA	Deoxyribonucleic Acid
ECG	Electrocardiogram
EMRI	Emergency Management and Research Institute
EPO	European Patent Office
ESRC	Economic and Social Research Council
EU	European Union

FDA	Food and Drug Administration
G2C	Government-to-Citizen
GDP	Gross Domestic Product
GEC	Group Executive Council
GEO	Group Executive Office
GP	General Practitioner
HLFPPT	Hindustan Latex Family Planning Promotion Trust
HR	Human Resources
I4M	Innovations for the Millions
IBRD	International Bank for Reconstruction and Development
ICT	Information and Communication Technology
IIMA	Indian Institute of Management Ahmedabad
IIT	Indian Institute of Technology
IP	Intellectual Property
IPR	Intellectual Property Right
ISRO	Indian Space Research Organisation
IT	Information Technology
JCI	Joint Commission International
JSS	Jan Swasthya Sahyog
km	kilometre
KSPCB	Karnataka State Pollution Control Board
KYC	Know Your Customer
LED	Light-emitting Diode
MBBS	Bachelor of Medicine and Bachelor of Surgery
MD	Doctor of Medicine
MFIN	Microfinance Institutions Network
MHUPA	Ministry of Housing and Urban Poverty Alleviation
MIT	Massachusetts Institute of Technology
MoUD	Ministry of Urban Development
MRP	Market-to-Retail Price
MSE	Micro and Small Enterprise
NCD	Non-communicable Disease
NEFT	National Electronic Fund Transfer
NGO	Non-governmental Organization
NH	Narayana Health
NHSRC	National Health Systems Resource Centre

NIH	National Institutes of Health
NITI Aayog	National Institution for Transforming India
NRHM	National Rural Health Mission
NSDC	National Skill Development Corporation
OLS	Ordinary Least Squares
P&G	Proctor & Gamble
PPP	Public–Private Partnership
PRADAN	Professional Assistance for Development Action
PRS	Passenger Reservation System
R&D	Research and Development
rDNA	Recombinant DNA
RMP	Registered Medical Practitioner
RSBY	Rashtriya Swasthya Bima Yojana
SCALE-UP	Slum Communities Achieving Liveable Environments with Urban Partners
SEARCH	Society for Education, Action, and Research in Community Health
SETU	Self Employment and Talent Utilization
SKU	Stock-keeping Unit
SNCU	Special Newborn Care Unit
STD PCO	Subscriber Trunk Dialling Public Call Office
TCS	Tata Consultancy Services
TGIFR	Tata Group Innovation Forum
TNMSC	Tamil Nadu Medical Services Corporation
TV	Television
UK	United Kingdom
UNICEF	United Nations Children's Fund
US	United States
USAID	United States Agency for International Development
VAT	Value-added Tax
WIPO	World Intellectual Property Organization
WTO	World Trade Organization

Introduction

Ancient Indian civilizations, particularly the Indus Valley civilization and the Mauryan Empire, have earned widespread recognition and fame for path-breaking innovations in spatial design and significant advances across diverse fields, such as mathematics, science, and astronomy, made during these periods. What is less known is the range and diversity of innovation in modern India. Since economic liberalization in 1991, there has been a surge of innovation in India, driven by entrepreneurs, corporations, and public–private partnerships (PPPs). Former president, Pratibha Patil, declared the decade of 2010–20 the 'Decade of Innovation', with a vision of innovation as the engine of inclusive economic growth for all Indians. The National Innovation Council, along with smaller state innovation councils and sector innovation councils, was created to encourage and support innovation across India's regions and industries. The most visible examples are in the category of 'frugal innovation', where a variety of products and services, available around the globe, are offered for a fraction of their cost and of a similar quality. There are other innovations as well, such as the use of economies of scale to drive down cost and develop entirely new business models for services such as wireless telephony. Another area of innovation includes new models of healthcare service and delivery developed for India that have spawned innovations in higher income developed economies. Finally, 2010 onwards, India has witnessed a surge in the technology start-up space, with more than 3,000 of them endeavouring to disrupt traditional models of business across areas as diverse as food delivery, airline ticketing, fashion retail, and competitive exam preparations.

There is no better time than now, therefore, to map the innovation space in India, and to understand the background conditions that have spurred on and, at the same time, hindered the innovative potential of a nation that accounts for more than a billion in human population. This book—a collection of narratives about innovation in India—aims to live up to this task. Since Independence, India has emerged as a vibrant economic power and a global leader in the international arena, and yet it continues to suffer from chronic problems such as inadequate infrastructure, overpopulation, environmental degradation, and mass poverty. In addition, India is a complex landscape with diverse customs and practices and little consensus on even basic needs and priorities. The bureaucratic red tape and inefficiencies that came along with the pervasive State presence from 1947 into the liberalization era of the early to mid-1990s have contributed, in no small measure, to the stultification of innovative potential. At the same time, with a burgeoning and young population—with a median age of 27 and with 30 per cent under 15 years—India presents a very large future market and workforce in all domains (National Commission on Population 2006: tables 17 and 18). In India, innovations are difficult to implement, yet all the more important because of their transformative potential. Overall, Indians have an optimistic view of innovation, but this is qualified by the real constraints that entrepreneurs, government, and leaders face as they lead India into the future. This book highlights these paradoxes and contradictions, identifying problems and offering solutions along the way.

The book, drawing from an interdisciplinary conference organized by the University of Pennsylvania,[1] examines the ways in which India addresses its challenges through innovation. Unlike existing literature in which innovation practices are narrowly defined within the business sector, this book provides a broader discussion of innovation by focusing on systemic and sectoral innovations spearheaded by a range of actors, from non-governmental organizations (NGOs) and firms to governments, courts, and grassroots communities. In short, this book focuses on innovation within the Indian society by all its significant actors. Its interdisciplinary approach offers a fresh perspective on innovation and the challenges and aspirational goals associated with it in developing

---

[1] India as a Pioneer of Innovation: Constraints and Opportunities, 14 and 15 November 2013, Penn Law, University of Pennsylvania. See https://casi.sas.upenn.edu/events/indiainnovation2013, last accessed on 15 March 2017.

societies. In addition, this book is the first of its kind on a topic that lends insight not only on India but also on innovation in low-resource settings more generally.

The book explores the broad story of innovation in India through a study of three themes: (a) innovation in business; (b) the legal and regulatory framework promoting and limiting innovation; and (c) innovation in public services and poverty reduction. These are perhaps the three most significant strokes from which one can derive a near-comprehensive account of India's innovation story, with all its challenges and potential. The first theme focuses on presenting the innovation story from the perspective of private actors who, motivated by both profit and principle, have hit upon and implemented novel and innovative ways to deliver products and services. The second theme addresses the conflicting values that innovation incentives bring to the forefront and the possibility of negotiating these challenges through legal and regulatory strategies. The final theme focuses on unlocking innovative potential to meet certain public ends, namely access to healthcare in rural and remote parts of India and urban poverty reduction. Through the various chapters, we get a stronger sense of how consequentialist approaches to innovation can shape the roles and responsibilities of multiple actors.

Those motivated by profit are traditionally considered to be amongst the top actors driving innovation, partly because such ordering fits well with the rational actor model employed in classical economics. While this ordering and its underlying assumptions may not be entirely borne out in practice, business is certainly a very significant source of, and platform for, innovative activity. This is the story the world over. Entities operating for profit innovate in order to enhance labour productivity, utilize raw materials more efficiently, reach out more expansively to consumers, and develop better products and services for consumption and, in return, maximize their profits. Global corporations such as General Motors, Microsoft, Sony, and Google are incubators of innovation driven by this conventional business model. India and her private corporations are no exception to this story. India's family-run corporations such as Tata, Birla, and Reliance are not famed innovators in the global industry research space when juxtaposed with Siemens, General Electric, Tesla, or Apple. But there are valuable lessons on innovation to be learnt from each of these traditionally structured organizations. They have all succeeded in a political climate resistant to private innovation by figuring out innovative ways and methods to cut costs and widen the market. We find many new-age Indian businesses that innovate using technology to solve problems unique to the Indian context. An interesting example is Consim Info, which managed to disrupt the marriage-broker model in India for arranged marriages through its various Web portals. Another is redBus, which identified the hassle of bus booking through private operators and innovatively resolved it through its online portal, reaping profits while at it. With the current political and economic environment favouring the growth of many more such start-ups, the future of business-led innovation looks more promising than ever before.

Indian innovators have also led the way in developing new, cost-effective ways to provide high-quality essential services. The Indian generic drug industry is one of the finest examples of this, providing life-saving drugs at very affordable rates to the populace. This strategy of catering to the volumes has worked effectively in other areas too, including healthcare and education. Narayana Health (NH), one of the case studies covered in this book, is a hospital chain with one of the largest networks for telemedicine. The chain started by offering low-cost, high-quality cardiac surgery to indigent populations. It then addressed questions of building an 'ecosystem' (including information, financial, and logistical infrastructure), in addition to performing surgeries for patients coming into the health system. The relentless pursuit of the joint goals of high quality (excellent post-operative results) and lower operating costs (via experience and process improvement) has resulted in a model superior to any in the world in terms of cost and comparable or better in outcomes. Another example is Aravind Eye Hospitals, which have innovated in cataract surgery by creating a manufacturing ecosystem at a lower cost, a well-functioning eye bank, and innovation in the process of cataract surgery that has dramatically reduced costs per patient while retaining very high rates of positive outcomes. A similar set of innovations has created a cellular telecommunications industry with amongst the lowest cost per minute of communication in the world, for over 900 million telephones.

Insights from the past are extremely helpful in understanding the present and the future of business innovation of such kind in India as discussed earlier. This book, therefore, begins with Claude Markovits' chapter, which offers a historical perspective on private innovation in India. Chapter 1 adopts a non-Schumpeterian conception of 'innovation', broadening the reach of this term to include non-technological innovations. After a brief survey of the existing literature, it seeks to identify patterns of innovation that were at work in the colonial period through two case studies: one of an industrial firm (Tata Steel); and the other of a merchant network (the Sindworkies of Hyderabad). It concludes with a brief review of post-Independence developments and critiques the stifling of innovation in this period, especially before India opted for liberalization and globalization in 1991.

With this background, Chapter 2 by Barbara Harriss-White proceeds to examine a very important segment in the Indian private entrepreneurship space: the 'informal' economy. It specifically enquires whether the fact that 9 of every 10 jobs and about two-thirds of the gross domestic product (GDP) arise from India's socially regulated 'informal' economy raises challenges to the evolution of innovative practices. Through her study of a South Indian town, Arni, and its professional and trade bodies, Harriss-White reveals the rich texture of innovative practices constantly deployed by both private actors and the state. In addition to identifying and understanding some of the innovative practices affecting daily life in Arni, the chapter also explores the roles played by institutional actors, including schools, banks, and business associations, in promoting rural innovation. This, in turn, throws light on the background conditions and support structures required for innovative practices to thrive in an economy shorn of the frills of a formal business enterprise.

While Chapter 2 disputes the prejudice against the possibility of innovation in informal settings, Chapter 3 shows how innovation can thrive even in low-cost settings. In this chapter, Tarun Khanna and Budhaditya Gupta scrutinize evidence emerging from a study of Narayana Health, the hospital chain mentioned earlier, to explore the large-scale public impact of private business in a field like healthcare. This chapter identifies some of the major institutional voids private enterprise faces in India due to the government's failure to put a healthy support structure in place. Since these institutional voids can choke off the flow of information, Narayana Health has had to invest heavily in alternate institutional structures. The insights from this chapter assume greater significance because of Narayana Health's foundational philosophy of providing affordable healthcare, especially surgeries, for the poor in India. The chapter identifies its challenges in relying on a low-cost model and shows how it innovatively overcame information gaps and other hurdles to deliver quality healthcare to patients in need.

Large business groups have always played a significant role in generating wealth and employment in India. Even after liberalization, a significant number of private businesses in India are managed and controlled by tightly knit family structures and closely held corporations—many of which are dominant players in their industries. Their need to, and routes to, innovation can result in drastic signalling effects as well as setting pioneering trends. In Chapter 4, Prashant Kale and Harbir Singh explore how innovation is becoming an important part of business strategies of many Indian business groups. Growing foreign competition and the desire to capitalize on growth opportunities at home and abroad have impelled them to plough resources into innovation. This chapter focuses on the mechanisms and practices that Indian business groups use to foster and promote innovation and entrepreneurship and compares their approaches to those used by multinationals in developed countries.

These four chapters, in addition to mapping the scope and terrain of business innovation in India, also demonstrate the significance of regulatory and support structures in determining the boundaries of the innovation space. Why is it that certain kinds of business innovations, in certain locations and in certain sectors, thrive while others perish? The response to this is closely tied with the narrative of governmental intervention and regulation in India, and the part-amicable, part-uneasy coexistence between business innovation and governmental regulation. This is a globally recurring theme. When governments stipulate hard and fast rules, rather than open standards, it disincentivizes business innovation in the rule-bound sector. When labour regulations are rigid, innovation of a more capital-intensive kind naturally follows. Liberal tax regimes attract higher capital flow. The Indian story on this theme is, however, unique, as Chapters 5 and 6 demonstrate.

Multiple questions arise when analysing the role of such interventions and its impact on the innovative abilities of the Indian public. Have state regulatory strategies ever been responsible for fostering indigenous innovation? Is the commonly accepted version of events—that of a post-Independence India resorting to a licence permit regime and thereby stifling innovation—true? Have the regulatory and legal shifts since the 1991 transition to a liberalized economy done any good in terms of fostering innovation? Chirantan Chatterjee and Shreekanth Mahendiran, in Chapter 5, use data from the pharmaceutical industry to address these issues and assess legal and regulatory structures and their role in mobilizing industry growth and advancing innovative practices. The chapter investigates, through an empirical lens, the emerging market for biological drugs in the Indian pharmaceutical market, using demand-side data for 22 biological large-molecule drugs across various Indian states, to understand the changing market structure. The authors follow their empirical analysis with managerial and policy recommendations.

Intellectual property rights (IPRs) are a significant area of incentives to innovation, one that can serve as an effective policy lever in the hands of the State to encourage or pipe down innovation on a general, or even sectoral, basis. In the Indian context, IPRs have also been the arena of conflict in policymaking due to the inconsistency between the domestic socialist philosophy and the internationally imposed overlay of free market thinking courtesy the World Trade Organization (WTO) and the Agreement on Trade-Related Intellectual Property Rights (TRIPS). The IPRs continue to demand attention as the centrepiece of the legal and regulatory structure for innovation in India, as amply witnessed by the policy measures and public pronouncements in this direction by the Narendra Modi government ever since it assumed power in 2014. The recent National IPR Policy, expressly targeted towards boosting innovation through IPRs and streamlining the ease of obtaining and enforcing these rights, is a landmark policy measure. For these reasons, studying the protection of IPRs in India is ideal in gauging the impact of governmental regulatory choices on innovation.

The property rights conferred on inventors by patent law and on content creators by copyright law play an instrumental role in incentivizing innovative activity. At the same time, the near-monopolistic character of these rights can end up limiting downstream invention and creativity. Chapter 6 examines these conflicts between incentives and access, within the context of copyright law. Shyamkrishna Balganesh and David Nimmer examine the difficult policy choices involved in resolving the conundrum of over-incentivization (through expansive copyright protection) and under-incentivization (through liberal exceptions to exclusive rights) in the content industry. They undertake a comparative study of two well-recognized approaches to crafting limitations and exceptions in copyright law: fair use and fair dealing. While fair use entails the use of an ex post facto standard-driven approach to examining the costs and benefits of a defendant's use, fair dealing usually entails a clear delineation of subject-specific limitations in a copyright statute. The authors show how Indian copyright law contains a mix of both approaches, which has allowed courts to expand the scope of exceptions when necessary and at the same time, unambiguously exempt certain kinds of uses from infringement. Contrary to common understanding, the chapter also shows how fair use can produce significant certainty over time, while fair dealing may, in turn, generate sufficient flexibility when the right set of institutional variables enters the picture.

Closely integrated with the understanding that business innovation cannot take off in the absence of a favourable regulatory climate—the insight from Chapters 5 and 6—is the recognition that there exist sectors where innovation has to be led by the State, or promoted with the state as an active participant and not a mere regulator. This may happen due to the inability of private actors to fully internalize the benefits flowing from their activity, thus disincentivizing them from looking any further in that direction. It may also be the case that private activity would merely be a drop in the ocean, because the problem that needs redressal can only be resolved with highly proactive State intervention of a scale and size unimaginable by private players. Explicitly recognizing the complex diversity that sets apart India from most other developing and developed nations also means acknowledging that no degree of private innovation can truly amount to much in India without the state and public sector adopting innovative practices in delivering public goods and services. Chapters 7, 8, and 9 are geared to throwing light on these themes. All of them focus on the 'public' and 'social' dimensions of innovation, thus elucidating *why* innovation is particularly important for the resolution of human problems and needs.

In Chapter 7, T. Sundararaman and Rajani Ved examine innovative efforts and endeavours in public health. The authors rightly argue that the trend so far has been to perceive problems in strengthening public health systems as inherent to their public nature and search only for market-based solutions. However, as shown in the experience and practice of the last seven years in India, much can be done with innovation in institutional design and building institutional capacity. The sources from which such innovations arise are many and unpredictable. However, there are ecosystems in which such innovations flourish and others in which they do not. Further innovations require rigorous evaluation and validation and considerable tweaking before they become ready for scaling up. The process of scaling up, too, is one of constant adaptation, not passive replication. To make these points, this chapter discusses the key successful (and not-so-successful) innovations in the Indian health sector with regard to PPPs, community processes, health information technologies, the quality of care, and human resource policies. It also mentions areas where innovation is required but has been slow merely because the ecosystem does not favour it.

Chapter 8, by Vijay Mahajan, addresses questions that go to the heart of this book. How do innovative practices pan out in low-resource settings, particularly when the population of the country in question is the second largest in the world and a significant segment of it is economically disadvantaged? What kind of innovations make maximal impact in such settings, and under what background conditions can such innovations flourish? The author explores the concept of 'innovation for the millions' (I4M) and its presence and operationalization within the Indian context. He examines 10 such innovations across the private sector, government, NGOs, and cooperatives, ranging from chicken hatching and milk production to mass-scale cataract operations, public toilets, and one-cent-a-minute mobile telephony. Mapping them, the author demonstrates—quite significantly—that innovations relevant to the millions are mostly spearheaded not by the private, for-profit sector but by other important actors in society: the public sector, civil society, and cooperatives. He attributes their success and impact to factors such as their demographically appropriate and socio-technical character, accurate identification of resource and reach gaps in the state's day-to-day functioning, and tremendous cost-cutting initiatives. While optimistic about the opportunities for the further evolution and scalability of such models in India, the author also cautions us about some constraints, the most important of which is the negative attitudes of the government and bureaucracies to such efforts.

In Chapter 9, Brian English studies some significant, recent, market-based solutions to urban poverty. The rapid increase in slum populations in India, now at 93 million, has far outpaced the impact of state interventions to date and necessitates a search for innovative market-based solutions. Market-based solutions have been gaining the attention of governments, international aid groups, NGOs, and entrepreneurs as they look for new ways to scale up their impact and sustain their interventions. This has given rise to thousands of new social enterprises that seek to provide social benefits and are sustained largely by business revenues. These initiatives are welcome news to governments and donors with tight budgets, and their rapid proliferation has attracted the attention of impact investors looking to create the next micro-finance industry. India has been a beacon of innovation in this field, creating enterprises that span from sanitation to education improvements. Chapter 9 profiles two case studies of market-based solutions that provide some important lessons for similar initiatives. These case studies demonstrate how partnerships between local government and social enterprises can deliver large-scale results, and show how important an enabling environment is for these supply-driven enterprises, in addition to evaluating, then fine-tuning, the business models.

These nine chapters are not meant to be exhaustive or to bring within their sweep every possible innovation currently being undertaken in rural and urban India. They also do not purport to predict the future course of innovation. Their message is about the highly unstructured growth of a country struggling to meet its needs, one that absolutely requires well-executed innovations in every possible sector. In this scenario, it is virtually impossible to provide a complete descriptive account or give normative guidance. This book, however, highlights some important trends and concerns. A common presence across each of the three themes—innovation in business, legal and regulatory challenges, and innovation in public services—is that of supplementary, rather than disruptive, innovation. Most successful innovations in post-liberalization India identify failings and gaps in the state machinery or ways to support already scaled-up businesses, and address them through the creative application of technology and manpower. Many innovative businesses are in the imitation game—especially in the technology start-up space—as they look to emulate models already proven in the West and adapt them to Indian conditions. Realizing and accepting this fact can go a long way towards suitably tailoring the legal and regulatory environment as well as the governmental response to innovation.

Reference

National Commission on Population. 2006. *Population Projections for India and States 2001–26*, Report of the Technical Group on Population Projections. New Delhi: National Commission on Population.

CLAUDE MARKOVITS

Historical Perspectives on Innovation in Indian Business

When it comes to business in India, placing innovation in historical perspective is a challenge. Existing theoretical frameworks
such as that of Joseph Schumpeter are of limited use. The literature on
the history of Indian business, apart from not being very substantial,
has paid little attention to the innovation question and, when it has, has
rendered unclear verdicts. In spite of that, the chapter starts with a brief
survey of the literature and then puts forward a few hypotheses based
on some case studies. In conclusion, it points to some of the remaining
problems.

A Brief Survey of the Literature

This section takes up the story only from the early modern period,
when European merchants and travellers provide us with testimonies
about the Indian merchant world, which, although not devoid of bias,
are still a major source for historians. The first perceptive observer of
Indian business we come across is Tomé Pires, who was apothecary
to King Manuel of Portugal and travelled widely in the East with the
Portuguese fleets in the early sixteenth century. He wrote a treatise
known as the *Suma Oriental* (1944), in which we find a description of

the trading world of the Indian Ocean and the South China Sea that remains unrivalled. Pires was struck by the scope of the network the Gujarati traders had built, which allowed them to dominate trade in that entire area. He was full of praise for the skill and enterprise of the Gujarati traders, whom he found in no way inferior to the best of Europe's merchants (Pires 1944).

When Pires wrote the treatise, the Gujaratis' rise to dominance in the Indian Ocean was a fairly recent phenomenon. It rested on both their skill as navigators—it was a Gujarati pilot who most probably guided Vasco da Gama on his voyage between Malindi in East Africa and Calicut (now Kozhikode) in 1498 (Subrahmanyam 1997)—and their unparalleled knowledge of markets and currencies in the entire region. There is no doubt that they were innovators in their time, although we know little about their financial techniques. In particular, we do not know if they already used *hundis*, the indigenous bills of exchange that would become such an important feature of the operations of Indian merchants in the following centuries.

While the first Europeans who traded in Asia thus recognized the advanced nature of South Asian trading practices, European views started to change in the seventeenth century with the arrival of the North European trading companies in Asia, primarily the Dutch East India Company and the English East India Company. These large capitalist organizations, the forerunners in many ways of today's multinational corporations (MNCs), could not immediately establish categorical dominance. They remained dependent for a long time, and in many ways, on local brokers through whose agency they could procure the goods they sought, such as spices, indigo, saltpetre, and, increasingly, textiles. However, the sheer range and scale of their operations—particularly the Dutch East India Company—as well as the military and naval forces they could muster gave them an edge over the local traders. The latter could not match their financial or military clout. In seventeenth-century Surat, then India's major port, there were a few Indian merchants, such as the famous Virji Vora, who held their own against the Dutch and the English, but with Surat's decline in the first half of the eighteenth century (Das Gupta 1974), the Europeans were left largely in control of large-scale maritime trade in the western Indian

Ocean. Some of these Europeans started feeling superior to, and dismissive of, the native traders and their acumen.

Two centuries later, in the twentieth century, some European trade historians saw the advent of the North European companies as akin to a revolution. The Dutch historian Jacobus van Leur (1955) put forward the thesis of Asian trade as 'peddling trade', inasmuch as it consisted mostly of a multitude of 'spot transactions' in which merchants themselves engaged physically and, because of the uncertainty inherent to such piecemeal transactions, differed fundamentally to the kind of contractual, large-scale transactions that were the mainstay of the European companies' trade. Enlarging on van Leur's insight, Danish historian Niels Steensgaard (1974) claimed that the European trade companies had fostered a real 'commercial revolution' in Asia in the seventeenth century. He noted that Asian merchants were slow in responding to the challenges of the time. For instance, they did not go for joint stock companies and limited responsibility until the mid-nineteenth century.

Whatever were the intentions of van Leur and Steensgaard, they created the notion (which became entrenched) that in Asia, Europeans were the innovators while Asian merchants were on the side of 'tradition'. For a long time, this dichotomy between 'modernity' and 'tradition' had a hold over even the best minds. It is very difficult, even today, to eschew that mindset completely and, when reflecting on the historical trajectory of the Indian business class, to avoid the implicit teleology of the superiority of Western capitalism. Some scholars, nevertheless, have attempted to challenge that viewpoint and provide a different kind of narrative. At a general level, there have been attempts to revise or at least revisit the question of the origins and course of the 'great divergence' between the West and the non-West. While Jack Goody (1996) and André Gunder Frank (1998) have been boldest in challenging the narrative of the 'triumph of the West', although not always in very empirically sound ways, Kenneth Pomeranz (2000) has offered a kind of mid-way solution by pushing forward the date of what he calls the 'great divergence' between Europe and Asia to the nineteenth century. Pomeranz's analysis was mostly focused on China; his treatment of India was weaker than that of Goody, who gave great prominence to the Indian case.

Two Indian authors with very different views command our attention. On the one hand, Irfan Habib (1990), the doyen of Indian Marxist historians, has challenged the thesis of the European-led Asian commercial revolution by stressing the fact that Indian merchants were institutional innovators to the same degree as Europeans. He identifies brokerage, deposit banking, bill money (hundis), and insurance as four areas of innovation and contends that Indian merchants' only significant weakness was their lack of military muscle as compared to the Europeans. Tirthankar Roy (2010) undermines the 'van Leur thesis' in a different way by emphasizing that Indian merchant communities were 'endogamous collectives' characterized by a high degree of internal regulation and, as a consequence, commercial disputes were not likely to arise, thus making spot transactions a perfectly rational way of conducting business. In the universe of Roy's endogamous collectives, a regulatory framework based on a law of contract appears totally superfluous. Roy goes on to indict the half-hearted introduction of European commercial law into India in the colonial era as having disrupted the functioning of collectives without ensuring proper resolution of disputes. This leads us to the question of the impact of colonialism on the Indian business world.

This is still treacherous ground, a kind of minefield in which political correctness is at times apt to trump scholarship. Two opposite views still hold sway, and they are difficult to reconcile. On the one hand, there is the view, which goes back to Mill and Marx, that the British introduced movement into a frozen structure and created some of the basic conditions for a capitalist transformation in India, even if these potentialities were not fully realized. The diagnosis can vary as to why. For an author like Morris D. Morris (1967), who came from the Marxist Left, it was mostly the nature of Indian society and the weight of traditional values and the caste system that prevented India from deriving all the benefits it could have derived from its link with the most advanced country of the time. For others of this view, such as Roy, the confusion engendered by conflating the two logics of contractual law and personal law (protecting the 'Hindu joint family') became the major obstacle to growth in colonial India. On the other hand, there is the standard Indian nationalist view—the 'deindustrialization' thesis—that the British impact on the economy was mostly destructive, particularly in its effect on manufacturing (Bagchi 1976). Both these views ignore the agency of businesspeople, looking as they do for causal factors of a structural kind.

If the lens shifts to view business motivations and attitudes towards innovation, especially of the kind injected by technology, a clearer picture emerges. For instance, why did the Indian textile industry, which until the mid-eighteenth century was the most advanced in the world (particularly in the quality of its dyes, which were much superior to those used in Europe), suddenly lose ground to the new mills of Lancashire? Does this not reveal a lack of innovation? The answer is that the industry had been innovative in adapting itself to new markets and different consumer tastes in Europe, but the artisans and organizing merchants saw no need for changing techniques that ensured a high level of sales. On the other hand, in Europe, technological progress appeared to be the only way to break the stranglehold the Indian producers had on the market. The technological conservatism of the Indian industry did not, at the time, appear irrational. Thus, the question of Indian entrepreneurship in the colonial era, to which we now turn, is essential to understand the trajectory of innovation in India.

Indian Entrepreneurship in the Colonial Era: Was it Innovative?

The literature on the topic is sparse and rarely very conclusive. In the 1950s, some scholars detected 'creative response' in certain sectors of the Indian economy. Dwijendra Tripathi (2004) and his colleagues at the Indian Institute of Management Ahmedabad enlarged and incorporated these insights into the general narrative of the history of Indian business. One could also mention Thomas Timberg's (1978) work on the Marwaris. While presenting a positive—at times glowing—view of the trajectory of Indian business, these scholars did not focus primarily on the question of innovation.

Before proceeding further, therefore, a brief discussion of my understanding of 'innovation' is in order. For Schumpeter, who remains the major reference on the question, innovation was at the heart of the dynamics of the capitalist system, in fact its only chance of survival (even if he was pessimistic about the long-term prospects of capitalism), and it was closely linked to technology. Actually, in Schumpeter's late work, innovation is de facto equated with technological innovation (McCraw 2007). A second important trait is that, for Schumpeter, the entrepreneur is not the capitalist, the owner of the means of production, but someone who is financed by a capitalist (Hagedoorn 1996), like Silicon Valley geeks are financed by venture capitalists. Now, holding to these two basic traits will make it difficult to identify innovation in the context of colonial India for two reasons: First, capitalist owners tended to be the only entrepreneurial figures around. Second, even they were rarely technological innovators. If we want to find innovation and not confine Indian businesspeople to the domain of 'tradition', as some would be happy to do, we have to go for a broader definition of what innovation entails. To illustrate what I have in mind, the following discussion offers two brief case studies: one of the best known of all Indian firms, the Tata Group; and the other of a lesser-known group of merchants I happen to have studied in detail.

Case Study A: Tata Steel

Tata Steel was founded in 1907 by the sons of Jamsetji Nusserwanji Tata, a major Indian industrialist who had been one of the pioneers of India's cotton textile industry (see Markovits 1996). Here, we focus on stressing three ways in which the creation of Tata Steel from 1907 to 1913 could be said to be a significant innovation in the Indian context of the time:

1. *Calling on the public to subscribe most of the capital of the venture*: This had never been done before; in fact, most of the capital for industrial ventures in India was normally raised from a limited milieu of merchants. It proved a resounding success, in spite of the risk involved. This has often been at least partly attributed to the atmosphere of nationalist fervour that swept India in the wake of the Swadeshi agitation of 1904 and 1905. Whatever the case, it was a bold move on the part of the Tatas and showed, for the first time, the potential of the Indian middle classes as investors.

2. *Combining in a novel way the different factors of production*: The Tatas erected their steel mills in the jungles of Bihar, in a spot equidistant from the iron and coal mines, thus minimizing transport costs. For their manpower, the mills drew on a combination of local labour, with 'tribals' mostly providing the unskilled part of the workforce, and labour imported from afar, either from other regions of India (Bengal for the clerks) or from abroad (Chinese carpenters and American and German technicians). It was a kind of cosmopolitan workplace, unlike any other in India at the time; and for a while, at least, the mix worked.

3. *Relying on the most advanced technology available at the time, mostly German and American*: The plans were designed by an American engineering firm, and Americans and Germans provided the core of the technicians. It was bold for an Indian firm at the time to dispense with British technology and technicians, who still dominated the Indian industrial landscape; that also is a testimony to the innovative character of this venture.

Case Study B: The Sindworkies

If Tata Steel represents undoubtedly the most conspicuous case of innovation in Indian business in the colonial period, other, more discrete forms of innovation can be detected by looking at the world of trade and finance. This case study focuses on the traders from Sind known as Sindworkies (Markovits 2000). Their innovation consisted in finding a niche that had remained unexploited and creating a worldwide trading network based in a provincial inland town. The original development was a remarkable case of serendipity: some merchants of Hyderabad, Sind, who had lost business in the wake of the province's annexation to British India in 1843, discovered that local craft productions sold well in Bombay (now Mumbai) amongst a European clientele to whom they became known as 'Sindwork' (hence the name 'Sindworkies'). The next step was that some merchants boarded ships from Bombay and started selling these products in Egypt around 1860, just as Egypt became a destination for wealthy European and North American tourists. From Egypt, some of these merchants ventured into the Mediterranean all the way to Gibraltar, then crossed the Atlantic via the Canary Islands to Central and South America. Others travelled from Calcutta (now Kolkata) to Singapore, the Dutch East Indies, and eventually China and Japan, where they started buying goods that gradually displaced the productions of Sind craftsmen in their sale catalogues. From such modest beginnings evolved a trading network that, by the early twentieth century, covered most of the globe, specializing in the sale of curios and silks to a mostly European clientele.

These Sind merchants innovated in two ways: First, they grasped that the growing taste of European middle and upper-class consumers for 'exotic' goods (particularly those produced in Japan, with the rise of *Japonisme*— the period of enamour with Japanese art—as an aesthetic trend) offered unlimited opportunities to traders willing to follow their customers in their travels around the world; and second, they successfully adapted a type of commercial firm developed by Europeans, with a central seat and branches, to the context of their small town. While the owners of the firms remained based in Hyderabad, they sent their employees to various places on two- to three-year contracts, taking advantage of the fact that Indians, being British subjects, could travel almost anywhere without documents, trusting to the protection of the British flag and Britain's unrivalled network of consulates. The capacity of traders from a small provincial town of British India to develop a durable worldwide network testified to their resourcefulness and gives the lie to the myth of the traditional, routine-bound nature of the Indian business world.

These two case studies have not been chosen at random. The aim is mostly to give an idea of the diversity of innovations that one could encounter in the Indian business world in the colonial period. I am not making a more general point about the innovativeness of Indian business; no doubt traditional routine behaviour was also prevalent in a big way. Yet, it is worth noting that this behaviour characterized British business in India as much as indigenous business and has, obviously, a lot to do with the overall trajectory of the Indian economy, which grew at a slow pace during the colonial period. The question is whether these 'green shoots' could grow in the more favourable context of post-1947 India. The next section attempts to open this question.

Innovation in Postcolonial India

The framework for business changed significantly after Independence in ways which, prior to the 1991 turn towards liberalism, were positive as well as negative. Innovation was not necessarily enhanced in the new setting, which often tended to reward a certain kind of opportunistic behaviour on the part of business owners. The impressive development of technological potential in postcolonial India did not necessarily translate into technological advances at the level of manufacturing. Investment in research and development was largely left to state institutions, with little involvement from the private sector. New industries were set up, but they generally made use of imported, sometimes obsolescent, technologies, like in the case of the automobile industry. Although new marketing techniques found their way into the country, the mom-and-pop stores remained the mainstay of everyday commerce. Finance too was not marked by much innovation. One significant novelty that began in the 1970s was the process of 'multinationalization' of some big Indian business firms—a largely forgotten forerunner of the present-day large-scale expansion of Indian firms abroad (Lall 1978). The period from 1947 to 1991, contrary to what could have been expected, was thus not marked by a great burst of innovation. As to the ways in which the scenario has changed after 1991, I leave to others in this book the task of enlightening us about it.

References

Bagchi, Amiya Kumar. 1976. 'De-industrialisation in Gangetic Bihar, 1809–1910', in Barun De (ed.), *Essays in Honour of Professor S.C. Sarkar*, pp. 499–522. Delhi: People's Publishing House.

Das Gupta, Ashin. 1974. *Indian Merchants and the Decline of Surat c. 1700–1750*. Wiesbaden, Germany: Fritz Steiner Verlag.

Frank, André Gunder. 1998. *Re-Orient: Global Economy in the Asian Age*. Berkeley: University of California Press.

Goody, Jack. 1996. *The East in the West*. Cambridge: Cambridge University Press.

Habib, Irfan. 1990. 'Merchant Communities in Pre-colonial India', in James Tracy (ed.), *The Rise of Merchant Empires: Long-distance Trade in the Early*

Modern World, 1350–1750, pp. 371–99. Cambridge: Cambridge University Press.

Hagedoorn, John. 1996. 'Innovation and Entrepreneurship: Schumpeter Revisited', *Industrial and Corporate Change*, 5(3): 883–96.

Lall, Sanjaya. 1978. 'Multinationals from India', in S. Lall, E. Chen, J. Katz, B. Kosacoff, and A. Villela (eds), *The New Multinationals: The Spread of Third World Enterprises*, pp. 21–87. Chichester, NY: John Wiley & Sons.

Markovits, Claude. 1996. 'The Tata Paradox', in Burton Stein and Sanjay Subrahmanyam (eds), *Institutions and Economic Change in South Asia*, pp. 237–48. New Delhi: Oxford University Press.

———. 2000. *The Global World of Indian Merchants c. 1750–1947: Traders of Sind from Bukhara to Panama*. Cambridge: Cambridge University Press.

McCraw, Thomas K. 2007. *Prophet of Innovation: Joseph Schumpeter and Creative Destruction*. Cambridge, MA: Belknap Press.

Morris, Morris D. 1967. 'Values as an Obstacle to Economic Growth in South Asia: An Historical Survey', *Journal of Economic History*, 27(4): 588–607.

Pires, Tomé. 1944. *The Suma Oriental of Tomé Pires*, Vol. I, edited by Armando Cortesao. London: Hakluyt Society.

Pomeranz, Kenneth. 2000. *The Great Divergence: China, Europe and the Making of the Modern World Economy*. Princeton, New Jersey: Princeton University Press.

Roy, Tirthankar. 2010. *Company of Kinsmen: Enterprise and Community in South Asian History 1700–1940*. New Delhi: Oxford University Press.

Steensgaard, Niels. 1974. *The Asian Trade Revolution of the Seventeenth Century: The East India Companies and the Decline of the Caravan Trade*. Chicago: University of Chicago Press.

Subrahmanyam, Sanjay. 1997. *The Career and Legend of Vasco da Gama*. Cambridge: Cambridge University Press.

Timberg, Thomas A. 1978. *The Marwaris: From Traders to Industrialists*. New Delhi: Vikas.

Tripathi, Dwijendra (ed.). 2004. *The Oxford History of Indian Business*. New Delhi: Oxford University Press.

van Leur, Jacobus C. 1955. *Indonesian Trade and Society: Essays in Asian Social and Economic History*. The Hague, the Netherlands: W. van Hoeve.

2

BARBARA HARRISS-WHITE[*]

Innovation in the Informal Economy of Mofussil India

Questions and Concepts

India's remarkable growth spurt—averaging 7.2 per cent between 2000 and 2008 (Denis et al. 2012)—has been called 'jobless' (Sen 2014). Jobs are in fact being created, but largely in the major sector of the economy that is 'unorganized' and unregistered (Harriss-White and Sinha 2007). This 'informal economy' is now linked directly through supply chains to consumption processes all over the globe. All stages of a production/distribution system may weave in and out of it.[1] In addition, many, if not most, firms in the informal sector are now selectively regulated:

[*] The field research, carried out with Gilbert Rodrigo, is part of a project investigating gaseous waste in the informal economy (White et al. 2012). This research has been funded by the British Economic and Social Research Council (ESRC) and Department for International Development (DFID), but the views expressed in this chapter are the author's own and not those of the funders.

[1] Forty per cent of India's manufactured exports are estimated to be produced there. For metal products, see Ruthven (2008) and for garments, see Messadri (2010).

licenced, but not paying taxes or complying with labour laws or environmental standards.

The term informal economy denotes a fuzzy concept with multiple interpretations and meanings (for example, small/primitive activity, unlicensed/unregistered, untaxed, and vulnerable workers without rights). Poverty, illiteracy, and other forms of deprivation reside there. For the most part, the informal economy also lies below the state's radar (Kanbur, Lahiri, and Svejnar 2012).[2] Illegal activity, which evades the law, overlaps with behaviour that pre-empts the law, or occurs in areas where regulations are not enforced or are neglected by the state and can be flouted with impunity. Informal activity overlaps with behaviour that does not come under the ambit of regulation at all, does not conform to it, or is developed prior to the imposition of regulation.[3] When the state attempts to enforce policies, its efforts typically lead to regulative distortions, extortion, and police action—to complicity, but not to compliance.

While India's growth has been widely attributed to the activities of the corporate sector in metropolitan cities (Sen 2014, citing DeLong [2003] and Rodrik and Subramanian [2004]), two-thirds of India's urban population of 285 million is located elsewhere—in mofussil (up-country) towns. These are frequently characterized as stagnant; Kamala Sharma (2009), for instance, evokes 'pitted roads, piles of garbage, open drains, stagnant pools of water, overhead electric wires'. Electricity, water, sanitation, and garbage collection are at best sporadic. But the contrast between metropolitan vibrancy and mofussil neglect is usually a matter of infrastructure provision, which points at the state's metropolitan

---

[2] Te Lintelo (2009) has analysed the 2009 National Policy on Urban Street Vendors as a means of evicting them on phytosanitary grounds. The Eleventh and Twelfth Five Year Plans have sections on 'inclusive cities', focusing on sites for street vendors, housing for informal settlements, integrating ragpickers into the municipal solid waste process chain, etc. In Mumbai and Bengaluru, planners are identifying specific zones for informal settlements and activity (Mohan and Rajagopal 2010).

[3] Labour, migration status, civil rights, tax, health and safety, land use, and environmental damage are increasingly prominent theatres of informality in the 'developed' economies of the West (Larsen 1992).

bias. India's towns are preserves of sociology and geography, but what about their economies? How innovative are the informal economies of mofussil India? This is the question that this chapter will address.

In Lundvall's (1992) authoritative review of the concept, innovation, like the informal economy, is a fuzzy term, but at the core of its multiple meanings and strategic vagueness, there is, at the least, novelty. For Lundvall, innovation is an interactive process, generating not only new products but also new processes and technologies, substituting new factors of production in a perhaps-unaltered finished product, a moment or process capable of changing the social relations of work, for example, through new labour requirements.[4] New forms of organization are innovations, such as the organizations and just-in-time practices of supply and exchange relations for the assembly of raw materials; the franchises, 'contractualized' organizations, and digital institutions; and practices which animate the processes of production and distribution and which control finance and labour. In new markets, two processes are at work, both of which involve innovation: (a) securing and defending a market share which may involve innovation in entry, in cost competition, in rises in productivity, and in labour displacement and (b) commodification, the conversion of things and activities into commodities for profit. New activities can create further markets for the product or service and give rise to other multipliers through invention and commodification (Leys and Harriss-White 2012). New kinds of persuasion create new forms of consumption and need. Kline and Rosenberg (1986) also point out that improvements in methods of innovation are themselves innovations, and to this we should add that the destruction and discarding of old practices is also a distinctive part of the process of innovation, one that is little examined in the literature. Innovation is often said to precede regulation (Dickson 1988), from which it is a short step to argue both that innovators resist regulation and that regulation stifles innovation.

Science and technology studies have developed the concept of 'innovation system': the regulative public and private sector institutions and information needed for an innovation to be (commercially) developed and diffused. Agricultural innovation systems (AISs), for instance, are

---

[4] Scholars of innovation often neglect the labour 'processes' themselves.

'systems of individuals, organisations and enterprises that bring new products, processes and forms of organisation into social and economic use to achieve food security, economic development and sustainable natural resource management' (Food and Agriculture Organization of the United Nations [FAO] 2012: 1). The AIS includes a multitude of potential actors such as producer organizations, research organizations, extension and advisory services, universities and educational bodies, governments and civil society organizations, coordinating bodies, individual farmers and farm labourers, and the private sector (including traders, processors, supermarkets, etc.) (International Bank for Reconstruction and Development [IBRD] 2012). This approach is useful because it allows for a great diversity of possibilities, consistent with real-world complexity. How relevant are these possibilities to the informal economy, however? Manufacturing and service provision in the informal economy have been recognized as continual, dynamic, active processes with indirect multiplier effects ('indirect network effects') and formal and informal interconnections throughout the system— from raw materials supplies to final effective demand (Roman 2008; Ruthven 2008).

A new family of concepts in corporate marketing addresses, to a certain extent, the question of innovation in India's informal economy. The term 'Indovation' encapsulates the idea that (Indian) poverty leads to creativity. The relevance of this insight to the informal economy cannot be doubted, though its implicit romanticization of poverty is another problem. However, the examples invoked—the Nano car, solar lights—are inappropriate, with origins and markets far removed from the abodes of the poor (Birtchnell 2013). Another concept is *jugaad*, a North Indian word for a cost-cutting, possibly unsafe quick fix, nearly always one that bends the law or ignores it completely. This is a peculiarly Indian contribution to the literature on 'frugal innovation', a powerful concept which is focused on multinational corporations (MNCs). Its allure, too, derives from its fuzziness. Frugal innovation celebrates least-cost ways to maximize profit, often involving an improvisational and flexible approach to innovation (rejecting research labs and innovation systems) and an exploration of the many sites inside a corporation where new ideas may originate. Consumers themselves, normally regarded as 'the market' outside the organization, may become further sites of frugal innovation. In this literature, the informal economy is the 'bottom of the pyramid', the market of poor, marginalized, excluded, un-bankable, sick and disabled, ignorant, or ageing consumers whose living conditions can be greatly improved by purchasing corporate innovations (see Radjou, Prabhu, and Ahuja 2012).

Proponents of the developmental desirability of jugaad and frugal innovation build their arguments from case material and inductive generalization. The cases selected in this literature are very distinctive—innovators are more or less educated people, capable of scaling up their novel ideas. These case studies, while not necessarily generalizable, provide a rich detail of 'lessons'—targeted at and driven by corporations. While frugal innovation will certainly have an impact on the informal economy (in many instances formalizing it), the entire family of concepts is geared reflexively to business, to strategic options inside big firms, rather than to an informal economy of tiny firms which expand by multiplication rather than accumulation and which are unaware of these concepts. We may indeed find frugal innovation in a literal sense in the informal economy, but we must distinguish its application in self-employment from the concept developed for the corporate sector, unless the latter appears in the informal economy. We will now examine it in the shape of an urban case study.

Method

The case study reported here is the economy of the town of Arni in northern Tamil Nadu, which has been studied between 1972 to the mid-1990s and almost continuously, though less systematically, throughout the twenty-first century—a uniquely long period (Basile 2013; Harriss 1991; Harriss-White 2003, 2016). Arni's census population has grown from about 33,000 in 1971 to well over 64,000 inhabitants in 2011, but the organic town actually exceeds 100,000 as it receives rural–urban migrants and engulfs villages outside its formal boundaries.

While there can be no 'representative town', this town has been accepted as a useful site to study rural–urban relations and local capitalism (Basile 2013; Jayaraj and Nagaraj 2006; Srinivasan 2010). It generates insights that we think are of wider relevance—and that may be tested by broader research.

Local business associations have long had a crucial role in regulating the urban economy, representing sectors, negotiating particular interests with the government, and controlling threats to their hegemony (Harriss 1981). Presidents of business associations are elected as knowledgeable representatives and, in past fieldwork, have proved able to tell sensitive stories from a comfortable 'third-person' perspective. In 2012, we discussed change and innovation with them. We interviewed a randomly selected 40 per cent of the presidents of Arni's 67 business and caste associations and major trade unions, last studied in 1997 along with 34 office bearers of salaried workers' unions (Basile and Harriss-White 2000).[5]

We had with us a note in Tamil and English, which sought verbal consent, and also explained our familiarity with Arni's past, the project's background, and our exploratory purposes. We also had a template questionnaire, which we learnt by heart and never showed to the respondents. In no interview was the full set of questions completed—an expected and routine feature of this kind of fieldwork. Establishing rapport with busy businesspeople requires allowing them to lead the encounter while the researcher attempts to structure it. This 'incomplete' method generates a database which provides the elements for analytical profiling and narrative and must be pieced together like a jigsaw puzzle.

The response was positive. Several interviews were prepared for in advance by the concerned presidents, who organized groups of up to 10 people to meet us (electricians, teachers, transport workers, and sanitary workers). These were not 'focus groups' but generated very wide-ranging and informative conversations. The total number of people to whom we talked about innovation and change was in the region of 75. The fieldwork took a month, with two seasoned researchers who have known the region for three to four decades.

The narratives provided a series of snapshots and micro-histories. While they were not comprehensive, they proved much richer than anticipated (and too many to write up as case studies).[6] We, therefore,

---

[5] The population of business associations may have expanded through the self-organization of firms dealing in commodities new to the town since 1997.

[6] See Flyvbjerg (2006) for an excellent discussion on the case study method.

rearranged instances of novelty and innovation and listed them by 'type' and 'themes'. Content analysis has generated an account which stresses types, processes, and institutions of innovation, rather than quantitative generalizations—leading to conclusions which engage with the research questions. We also absorbed quotations into the narrative to illustrate the comments men (and some women) in Arni made as they struggled to make sense of their conditions.

The Town

> 'Arni is not a poor town ... Arni's economy is in good shape.'
> —President of Arni's Chamber of Commerce, June 2012

Arni has its own distinctive history and geography. The spatial and power configurations of its economy have been reconfigured and relocated a number of times since Independence. Then, the heart of the bazaar was located in what is now an impassably narrow, pig- and rat-infested side street just wide enough for a single van. The weekly market site near the old centre no longer exists, having been transformed into a bus stand; its former regional role in the marketing of cattle has also disappeared, with the replacement of animal traction by fossil fuel (Gathorne-Hardy 2013). The pre-Independence site of the bazaar is still the heart of the congested informal business economy. To the west, the cluttered bypass to this throbbing heart, incongruously named Gandhi Market Road (where statues of Mahatma Gandhi and B.R. Ambedkar vie for public space), has been a central business district since the 1960s. But, in 2011, the road was reinforced and widened and a motorway-style central barrier was erected which prevents customers from crossing it. The wide earthen verges were paved over, making it difficult to accommodate the cycles, motorbikes, and squatter stalls that once lined it.

Though it is often filthy and there are serious infrastructural challenges, the economy of the town is not run-down. New residential quarters sprawl to the west, while the relatively neat silk quarter, a product of the 1950s and 1960s, occupies the south of the town, where dyeing, spinning, weaving, and marketing compete for space with residential housing. Rice godowns (warehouses) used to line a number of streets in the heart of town but are now located on the periphery, where their owners profit from lower land values and rents. A new bypass encircling the entire town is ringed by fully automatic rice mills. North of the newest bypass—but umbilically connected to the town through roads and transport, residential demand, and low-order service jobs—a huge tract of agricultural land has been turned into a private higher education cluster by a single investor—a politician who belongs to the political party in power since 2011 in Tamil Nadu, the All India Anna Dravida Munnetra Kazhagam (AIADMK).

The town's economic base has been agricultural marketing, general retail, energy retail, and administration, together with a small industrial district for silk handloom weaving and a cluster of goldsmiths and pawnbrokers. Today, the bazaar economy of more-or-less independent, small family businesses is not only being supplemented by the private education hub but also threatened by a new scale of subnational and national capital, with branch firms and agents for sectors such as cement, dairy products, fuel oil, and supermarkets.

The town's engagement with the global economy is, at best, indirect. It is linked through the export of silk products to Southeast Asia, and even China, through wholesale firms in Chennai, the export of fodder from rice mill byproducts via Chennai to Europe, and the import of clothing (and even laminated wooden furniture) from China. The local workforce is drawn into global value chains. Every day, vehicles from Nokia and other factories on the Chennai–Bangalore corridor to the north-east and from the leather-export factories around Vellore to the north-west scoop up a (semi-)educated labour force from the town and its hinterland, enabling workers to commute quite long distances for work. For the rest, Arni is being globalized by pre-emptive responses to threats, rumours, and images, rather than by the active intervention of global supply chains. Metropolitan supermarkets are rumoured to threaten the town by establishing branches, so small firms develop supermarket layouts, personalize brands, and abandon haggling. Images of consumer society drench the television (TV) channels and create desires that local entrepreneurs turn to concrete supplies—notably, for instance, in women's fashion and beauty products.

Does this informal economy innovate? If so, what kinds of innovations emerge from informal institutions? Our analytical narrative focuses on examples that most richly illustrate the political economy of three kinds of innovation: invention; adaptive innovation; and adoptive innovation.

Kinds of Innovation in Arni

Invention

The first type of innovations we discuss here are those invented in and around the town. Though some of these innovations were also invented elsewhere, Arni did not adapt or adopt them from outside but developed and implemented them through internally driven acts and processes.

We will celebrate the electricians of Arni. The town's electrician workforce has grown from only 85 in 2000 to about 730, of whom 320 are registered. Only 20 of these have any formal qualifications, with many others having inherited licences from their relatives. Most electricians are 'undereducated', low-caste village men who learn on the job while they work for self-employed seniors, and all of them are keen to experiment. Arni's electricians' *sangam*, or association, has started to issue certificates based on experience alone. Being an electrician is a physically dangerous occupation, and the association has a strong esprit de corps. 'The work we do is not illegal, but it is informal', explained the association's president.

Faced with a chronically inadequate and unreliable supply of electrical power for irrigation pumpsets and a constant demand for equipment repairs, Arni's electricians have succeeded in modifying a three-phase technology to a two-phase power supply.[7] This enables pumpsets to ride out oscillations in power, rather than breaking down.[8] The invention involves adapting a condenser (a device to store energy and release it in spurts) by switching fuses, which aids efficiency. This device is cheap: in 2012, it costed INR 600–INR 1,000 to purchase and install it and it is reckoned to last two to five years. Electricians continue to work on new

---

[7] Three-phase electricity comes in three separate streams (for which three separate wires are needed), designed to oscillate in sequence to generate power.

[8] This is not to argue that this invention does not exist elsewhere, but the Arni electricians do not know of a parallel invention. See Shah and Verma (2008) for a similar case in Gujarat.

ways of compensating for irregular power and sudden total power cuts through battery inverters and voltage stabilization.

There is some dispute about the inventor ('was it Anandan or Annamalai?') and the invention's date is only vaguely remembered, but as the invention yields no royalties, the individual entrepreneurship is not as important as the invention itself. The electricians describe themselves as 'open', constantly learning, passing on knowledge, and experimenting on the job.[9] The risk of on-site experiments is explained to the owners on whose equipment the work takes place: 'If experiments fail, we repair the failure', said one unregistered electrician.

It was independently corroborated that the Arni invention has been formally scaled up and developed with an automatic switch application that is now being mass produced by a Bengaluru engineering firm 'after consulting' Arni electricians. Not to be outdone, the electricians are adapting the mass-produced product even further.

In sum, in this case, a growth in numbers together with increasing informalization (lack of registration) coexists with informal formalization (certification). The state's failures to enforce standards in supplies, in its own utilities, or in consumers' appliances, and to supply power without cuts or voltage fluctuations acted as triggers, creating economic problems that have generated entrepreneurial responses. Property rights are irrelevant to the stages of development and diffusion, but scaling up was achieved through free transfer and private appropriation by a company with access to finance and marketing. Experimenting on customers' appliances became the informal equivalent of the research lab, and continual interaction between a formally registered firm and informal labour enables diffusion (albeit with a large difference between the two in returns).

Adaptive Innovation

Tucked away on an upper floor of a poorly constructed building in Arni, in a narrow alley lined with open drains, is a 'computer centre'

---

[9] For an example at one extreme, a bottled cow dung paste is used to help motors start with two-phase power; at the other extreme is the use of electronics to operate small power generators by remote control.

established in 2010 in which innovations have been cleverly adapted to the local economy and society. The 'centre'—a room with a balcony— is collectively rented in a complex way by five young graduates, two of whom manage the centre for the other micro-investors. They work part time there and part time at salaried work in another town some distance away. They collectively own 10 desktop computers. Its stated objective is 'software training': it issues certificates to those who pass a three-month course in the rudiments of computer-aided design (CAD) and 3-D design. The clientele is village based and too poor to pay the INR 3,000 fee as a lump sum, so the introduction of a trust-based instalment payment, receding into the future well after the end of the course, has increased the centre's social reach. There is an empowering social benefit as a byproduct: through this training, Tamil-speaking students are forced to acquire basic computer literacy in English.

This is a novel, adaptive, organizational response to lack of adequate capital and adequate demand, in which formal educational skills are diffused into the informal Tamil-language economy in return for an informally formalized qualification through certification.

Adoptive Innovation

Adoption is by far the most common route to innovation. It is what local people understand as innovation, and the rest of the discussion of field material is devoted to the derivative spread of technologies, products, and practices invented and developed elsewhere. To the argument that adoption is not innovation, we respond that, first, all new activities require ingenuity and risk-taking with capital, and second, what is novel is socially embedded. In 1973, my interview with the first Dalit rice retailer in Vellore's well-established municipal market revealed that he faced huge social obstacles and knew himself to be in the vanguard of innovation for his caste.

Adoption is not new. Analytically framed in terms of innovation, the Green Revolution of the 1960s epitomized rural development in the form of continuous adoptive innovation and creative destruction (Harriss 1972). In the local non-agricultural economy, formerly un-commodified or semi-craft commodities or practices have continuously disappeared, to be replaced by new mass-produced commodities (crushed seashells by paints, turmeric by cosmetics, dhobi services by dry cleaning, and open defecation—to some extent—by lavatories with septic tanks). On a micro scale, this process has combined capital bias and labour displacement with caste continuity, resulting in economic differentiation within castes. But what our 2012 field evidence recorded is happening at a revolutionary scale, pace, and scope never encountered before. It involves the following:

1. *New products*: Not simply the ubiquitous mobile phone but also the form, quality, and symbolic meaning of basic foods, including dairy products, ready-mades, and alcohol. For example, the long-standing association between properties of food and health (hot and cold, pure and impure, vegetarian and non-vegetarian) has long been superseded by modern medical reasoning, in which science is yoked to advertising. This now pervades the informal food economy of Arni: high-quality parboiled rice is marketed for nutritional and health benefits, not only to the metropolis but also locally, accompanied by tenacious advertising.

2. *New processes*: The most prevalent innovation in the bazaar is new models of price formation: out with haggling (for old irrigation ironware, for example) and in with fixed prices associated with branding and packaging, even for activities not regulated by the state (such as unregulated new hardware products and interior decoration items). New process technology emphasizes the importance of infrastructure, the unreliability of which may constrain adoptive innovation associated with the commodification of processes: deep-freezer cold-chains for industrially processed milk products are abnormally disrupted by power cuts, as are electric sewing machines. Such innovations may even be de-adopted.

3. *New services*: Formally registered bank branches, automated teller machines (ATMs), and national electronic fund transfers (NEFTs) have revolutionized market transactions and helped to destroy informal credit relations. Subsequent, informal innovation using bank loans may be benign (investing in takeaway food services from 'meals hotels', for example) but not necessarily so (as in the adulteration of liquor). Formal innovation may also destroy informal

commodity economy (for example, the disappearance of toddy and arrack production due to its replacement by Indian-made foreign liquor (IMFL), branded beers, and prestigiously labelled alcopops).

4. *New forms of exchange*: The banking sector's payment revolution has been of paramount importance in the destruction of personalized exchange relations, with widespread institutional ramifications. Amid the proliferation of new informal financial institutions (Polzin 2007), one such ramification is the role of bank loans in the emergence of the regulated market site for grain transactions, after decades of avoiding loans (since production debts were repaid in kind at sites specified by the trader-lender). The first transaction between producer and trader is now formalized through the open auction of paddy, with payment through the banking system. A new registered and organized labour force works there too. Old informal aggregators—such as lending to marginal farmers to bulk up their small surpluses—have found new niches in the formalized system, in which they pass themselves off as primary producers. Another ramification of new exchange possibilities is the facilitation of pan-Indian supplies in groceries and foods and international flows in textiles—all brokered through known intermediaries but from unknown suppliers, using NEFTs. The removal of restrictions on bank loans for pure trade has permitted new scales of finance for working capital, such as 'a crore in a store' for rice mills, some of which are insuring their stocks for the first time.

5. *New ways of knowing*: The transformative and knowledge-empowering properties of the mobile phone have led to its universal diffusion such that, by 2012, few regarded it as an innovation. But whereas, in the 1990s, access to the Internet was the closely guarded domain of the business elite, it is now easy to obtain via mobile phones. Advertising to create and bind loyalty and avoid competition has become macro audacious and micro precise. Even in the informal economy of weaving, the manufacture of image is vital to marketing, and also to production. In silk, relabelling Arni's industry as that of Kancheepuram—where higher-quality saris are produced nearby—has had a significant impact on processes, products, and sources of innovation. It is as much by employers' branding entrepreneurship as by the innovative behaviour of the craft-producing workforce that the threat of declining demand

to the industrial district for handloom weaving (at its peak involving thousands of looms) has been turned into the stimulus of a higher-status market. Conversely, the town has been unbranded for rice. While in the 1990s 'Arni rice' was a brand sufficient to frame this high-quality parboiled commodity, now a single rice mill typically develops and markets three independent brands.

We now turn from scanning the productive base to examining innovation in institutions and organizations.

Innovation by Labour

An unprecedented exit by semi-skilled and skilled labour has produced much innovation in informal labour processes, including mechanization, feminization, and migration. Local employers also concede demands for the 'informal formalization' of labour-force skills, which they know will reduce workers' dependence on them, depersonalize market transactions, and, in the case of men, lead to migration out of Arni. Employers then respond innovatively to in-migration, for example, with language training and increased surveillance. Labour markets have grown increasingly segmented by the precision of tasks. New levels of complexity in the labour market involve contradictory as well as complementary social processes: specialization fuels an expansion of self-employment, while the biggest firms do the opposite and 'vertically integrate' skilled labour (for example, in construction and rice milling); it is the cheapest, not the costliest, labour that is displaced by productivity-raising technology, the removal of workforces tainted by caste, gender, and rurality being more socially urgent than the impact of capital-intensive technology on the costs of production.

In the process of informal innovation, however, we found sharp contrasts between collegial support for invention, adaptive innovation, and adoption on the part of petty artisans and small capitalist firms on the one hand and the oppressive conditions of low-caste wage workers, which are not conducive to innovation on the other hand. In the case of transport workers, these conditions included delayed payments, partial payments, refusal to renegotiate fixed wages in the face of inflation, refusal to employ on permanent contract, etc.

Changes to Family Businesses

'Men don't want to work for others', the president of the tailors' association told us. 'Self-employment is an aspirational state'. Under the Indian Constitution, the private sphere, or the family, is regulated by a combination of statutory law for the Hindus and customary law for the plurality of religious minorities. In self-employment as well as the family firm, however, the private meets the public. The customarily regulated family is the building block of the economy. Collective property rights and informality are therefore hardwired into it.

In Arni, there has been a massive expansion of small firms run by undereducated, self-employed people. They are associated with many of the above-mentioned process and product innovations, such as the electricians and pumpset retailers and mechanics (who went from 10 shops in 2000 to 70 in 2012). The new skills needed for such work are 'learnt by doing'. Throughout India, this expansion of modern artisanal activity by multiplication of firms, rather than by the concentration of capital, is an uncelebrated and poorly explained feature of liberalization (Harriss-White 2012, 2014).

At the same time, firms undergo specialization and complexification. Even tailoring has reorganized itself to handle demand through 'vertical disintegration' on a micro scale, with separate 'firms' specializing in repairing, cutting, sewing, and pressing and the newly gendered segregation of female tailors for women's clothes.

Family business is also being ruptured and reorganized. In the central bazaar, the mass exit of the English-educated younger generation to salaried jobs in the nearby information technology (IT), electronics, and auto industries is causing a decline in family labour. This is compensated for by an increase in wage labour. 'Only scions with "business-level education" [low education] stay in the bazaar economy—it's hard work, long hours, competition, no holidays—with the further problems of apprenticeships', a pumpset dealer told us. The entry of wage labour into tiny firms reinforces the shift in the 'crafts and craftiness' of the bazaar, in which trust-based haggling over price, credit, and debt (hitherto the work of junior family members) is replaced by norms and electronic exchange. The modernization of exchange and the extinction of the multi-generation petty business is likely to be a concatenation event involving many innovations being absorbed simultaneously.

An Innovation System? The Social Structure of Informal Innovation

Social institutions combine to form a structure which is a puzzle in social science. For new institutional economists, institutions are constraints on (profit-maximizing) activity. For old institutionalists, they make economic activity possible in the first place: for the 'social structure of accumulation' school, they form a matrix stabilizing capitalist accumulation and controlling conflict and for Marxist institutionalists, they express forms of economic and social authority which detract from class formation. All these approaches differ with respect to the institutions considered to be indispensable, with some arguing that this is an empirical question, others using the circuits of production and reproduction as a guide to what is most important, and yet others using ideas of dominance and prominence or the concept of pre-capitalist institutional 'outliers'.[10] In the absence of anything resembling a formal innovation system in the informal economy, the questions at issue here are empirical ones: How do institutions nurture informal innovation? Which institutions form obstacles? Do these institutions form a structure? This tentative attempt to answer them from a pilot field project is naturally contestable. Here, we tie the evidence from the case material into a higher-order narrative.

Over and above the threats to, and transformations of, the family firm, three other kinds of institutions stand out for the frequency with which our interviewees mentioned them: schools, banks, and business associations (the latter is hardly surprising since we interviewed their

[10] See Harriss-White (2011) for a fully referenced review. E.O. Wright's 'predisposing factors' for innovative risk-taking in worker-managed firms, for instance, include credit markets (subsidized interest), stable income flows to labour, associational democratic governance, risk pooling, geographical anchors, and the known existence of technologies lowering diseconomies of small scale (Eric Olin Wright, personal communication, 2013).

presidents).[11] All are formally registered, but they result in informal innovation. All are interpenetrated by 'informal' institutions of identity (caste, ethnicity, religion, and gender), all of which mould life chances and people's capacities to act in the economy.

Institutions of education form a diverse ecosystem involving state, registered private, and informal private ownership (including tuition centres, apprenticeships, informal training centres, learning by doing, and absorbing media messages). A high level of formal education facilitates the evaluation of information (from the media) and access to supportive institutions (like banks). 'Arni has increasingly educated people who read adverts, watch TV, and are influenced by the media', said the chairman of the chamber of commerce. Education and/or competence in an informal knowledge base (such as weaving, electrical repair, or vehicle repair) provide the continuity that enables the adoption of new practices and other kinds of agency in which material technology and behavioural norms are transformed. The lack of these capabilities hampers innovative agency.

Formal education most resembles an innovation system, but one where formal and informal knowledge institutions intertwine. While some kinds of innovation need training, skills, and education, and while English is a passport to economic mobility, education generates its own paradoxes. Much knowledge is learnt on the job in various ways, and through self-skilling. People with low levels of education are not un-innovative.[12] People with high levels of education are not necessarily given to innovation or required to innovate inside complex organizations.

Education is, however, deeply embedded in 'institutions of identity', and their intersection affects agency and innovation. While women are well represented amongst classes of workers enduring chronically oppressive conditions and have entered the labour market as homeworkers, no

---

[11] See Harriss-White and Rodrigo (2013) for a more extended discussion of the contradictions of the state's involvement in the informal economy, which is out of the scope of this chapter.

[12] Indeed, Anil Gupta (1999) has characterized the poor as knowledge rich and devoted his life to recording the vitality of innovations by poor people in agriculture and rural development.

woman was mentioned as an entrepreneur. (Educated girls and women aspire to salaried jobs, an educated groom, and reduced dowries.)

Modestly educated Dalits work in occupations not requiring education: fruit and vegetable marketing, sanitation and recycling, portering, 'rooftop work', blacksmithing, lorry driving, and, lately, informal finance. Though one Dalit is the administrator of Arni's municipal bureaucracy and another is a high-ranking teacher, the well-known local role models for Dalits are three illegal moneylenders with large houses and swimming pools (an innovation in Arni), whose financial careers emerged from portering and fruit- and vegetable-selling and from street wisdom rather than formal education.[13] Other Dalit aspirations for upward mobility are different from those mentioned by the higher castes: the police, the army, and chauffeuring. For the minorities (Muslims and Christians), new reservations (at 3.5 per cent each) have been carved out in state employment. Muslim children are educated in Urdu until the sixth standard, after which they go to government Tamil schools, where they are reported 'not always to perform well' (according to one teacher). Muslim girls are also starting to obtain education but 'do not transform it into work afterwards' (another teacher). The lower-status identity groups, in encountering social obstacles to the exercise of agency, experience formal education as necessary but far from sufficient.

Despite the town being a 'low-caste place', it is clear that being a Dalit, belonging to a religious minority, or being a woman confine work possibilities for roughly two-thirds of Arni's population to positions where it is much harder to innovate in the informal marketplace.[14] The aspirations unleashed through education, and often thwarted in the marketplace, are not only expressed through changes in norms and economic mobility but also through changes in motive and behaviour. Although profit and a higher standard of living are reported as proof of the benefits of adopting new products and processes, other, less-obvious economic motives, such as collegiality, the expression of social status,

---

[13] They are said to lend at interest rates of 5–20 per cent per week and occasionally, 10 per cent per day.

[14] Muslims and Christians are about 10 per cent of the population, Dalits are about 15 per cent, and women are 50 per cent.

and the desire to nurture talent in others, are evident when they clash with profit and still prevail.

The effects of the influx of formally registered national banks can hardly be overestimated. First, these banks encourage innovation in new scales of technology, with subsidized loans encouraging capital bias, which in turn requires high rates of capacity utilization. Banks now also allow working capital loans for pure commerce (forbidden until the end of the twentieth century). There is no monitoring of virement or onward lending into the informal money markets: 'As long as we repay the instalments, the bank doesn't interfere at all', said a leading rice miller.

By virtue of their collateral requirements (such as title deeds), banks may play a role in formalizing the informal economy along the lines advocated so influentially by de Soto (2000). Loans for education and housing have rapidly increased in size and frequency. Formal property may secure (multiple) loans from (multiple) formal accounts that are subsequently invested in the informal economy or lent onwards to others. 'These are impossible for banks to monitor', explained a bank manager. Titles may be vested in a collective (the family) under the customary laws and hard to associate exclusively with an individual. Additionally, 'there have been problems with duplicate [forged] title deeds in Arni', said a banker. Certain banks do not require collateral to lend to some occupation groups: tailors mentioned their ease of access to formal loans which had been altogether denied to them 10 years earlier. Dalit sanitary workers in receipt of computerized salary transfers may be awarded loans of up to 10 times their salaries and are now accumulating formal debt for the first time ever. Yet, the new banks are far from vanquishing the big informal financiers in Arni, who remain businessmen (traders and agents), landlords, and 'finance corporations' (in which a group of savers, including government employees, invest).

Innovation requires risk-taking. For electricians and construction workers, this risk may be physical, but for the most part it is financial. We have no direct evidence about the equivalent of venture capital in the informal economy but, since no one mentioned it, we conclude that its lack is a constraint on disruptive innovation (as noted earlier, the scaling up of the electricians' invention was taken over by a large registered company in Bengaluru). Innovation through adoption and the exercise of agency is far less risky in monetary terms. Nonetheless, risks—environmental, health, financial, and political—permeate the informal economy. 'Half the town is saving against shocks', said the president of the Rotary Club. Sickness, marriage alliances (despite the downward drift and increasing optionality of dowries), private school, and college education can be as 'shocking' to the budget of a small firm as weather-related downturns or business losses. For shock-absorption purposes, bank interest is not high enough to compete with returns from property and gold, which can be cashed or mortgaged easily in town. In response to the demand for gold, a large jewellery company from Chennai is establishing a store in Arni. Yet, as with informal moneylending, the biggest savers are widely said to be government officials, who have the least shock-prone work conditions and contracts.

Beyond the state's reach, informal economic activity is regulated either as though the law were being enforced (Olsen and Morgan 2010) or through the decisions of business associations. These organizations span the entire spectrum, from the formal and nationally federated to the local and unregistered. Basile (2013) studied them closely in 1997.[15] She sees caste as a crucial element in a corporatist system of economic regulation, in which the ideology of caste[16] is secularized and increasingly internally differentiated, while the institutions of caste move from the domain of culture to that of the economy and are more or less mapped onto business associations. These are 'manifestations of the social order': defining behaviour, co-opting members across classes, and thwarting the development of class consciousness. Their regulative roles may include screening entry, apprenticeship, defining contractual measures and types, credit, price control in the 'market'—especially for labour and derived markets such as portering and transport[17]—working

---

[15] Basile uses the lens of Gramsci's theory of hegemony, in which the economic interests of capitalism use non-economic, political, and cultural means to co-opt subaltern classes.

[16] Caste is defined by Basile (2013) as a hierarchy of social status associated increasingly loosely with occupations.

[17] Arni's silk association has a long history of state-connived containment of informal wages for weavers.

conditions, the guarantee of livelihoods, and mobilization to compensate for accidents and premature death, poverty, and social distress. Basile stresses the importance of the state in business association's activity: the need to limit the state's intrusiveness (for example, from the Labour Act, the Packaging of Commodities Act, and value-added tax [VAT][18]); to protect members from the police; and to appeal against discrimination, while campaigning (often using bribes) for preference, infrastructure, and social rights and contracts.

Two questions arise: How have business associations themselves innovated since the 1997 research? What role do they play in informal innovation?

Many associations exist reactively, spurred into action when threatened by the state or when needed by members. A few (such as those for rice, silk cloth, groceries, and gold) have real clout, enhanced by federation and political influence at the ministerial level, and can boast of achievements such as reduced power cuts for rice mills and an agreement to let women enter tailoring. But federation is just one innovation that has gathered strength over the last 15 years in Arni. A second is support of business associations for informal training and accreditation for skills, a human 'collateral', which, in turn, permits the development of depersonalized transactions and migration for work. A third is the circulation of trade information, useful for the adoption of innovations from elsewhere. A fourth is the emerging cosmopolitan character of membership, weakening the alignment between caste and trade associations and replacing ascribed merit with acquired skill. The exchange with the state, itself fully capable of innovation, includes redefining the boundary between formal and informal activity (education, energy), selective enforcement (licences), tolerating informal formalization (certification), and neglect. The state's neglect of its regulative obligations results variously from the capture of policy processes by interests that would be curtailed were such policies implemented as apparently intended in law. Alternatively, regulative neglect is caused by scarcity of

[18] The state requires published maximum retail price indications, inventory lists, certified weights and measures, and quality control in retail, none of which was being observed in Arni. This was resolved when the state conceded to an agreement for incremental and delayed implementation.

personnel and/or equipment (as in the case of the district's Pollution Control Board manned by one official without a vehicle, or the schools inspector with no transport or laptop). Business associations and guilds grounded in caste are gatekeepers for entry as well as screeners of innovation. In a politics of representation, regulation, and mediation, this set of corporatist interests negotiates collegiality within a sector just as it controls the erosion of social barriers to entry.

The combination of education, banks, and business associations does not yet form a 'coherent' structure of innovation. The informal economy equivalent of the innovation system modelled in science and technology studies spans the bounds of formal and informal regulation in idiosyncratic ways and seems to operate with few interlinkages— even informal ones. Alternatively, laws, regulations, and administrative procedures, that range from amorphous to overspecified, interact to trigger innovation in ways that need new, granular research.

***

In India, 95 per cent of enterprises and 92 per cent of jobs are in the informal economy. So, the question whether the informal economy is the abode of enterprise is an important one, though little examined in the literature. The field research reported here confirms that India's macro-level growth spurt is the result of an explosion of innovative activity, agency, and institutional churning at the micro level in the informal economy. This research has revealed examples of:

1. Schumpeterian entrepreneurship—the inventive bridging of factors of production to create new production methods and technologies in commodification, cost competition, products, and forms of organization;
2. adaptive technological and organizational innovation;
3. transfer and diffusion of known technologies, processes, products, and organization;
4. diffusion of knowledge through new roles for the media and education, new images just as capable of transforming local production relations as of feeding new social and economic aspirations;
5. other motives for innovation than the realization of social aspiration and status (not only profit, livelihood, and the standard of

living but also the problem-solving mentality known as jugaad; the need to respond to selective state failures in provision, regulation, and enforcement; preemptive reactions to threatened change; and collegiality / collective advancement and nurture);

6. innovation and agency by and for labour as well as by and for capital;

7. radical, disruptive innovation (banks' NEFT, threats to the small family firm) as well as incremental innovation (the cases of informal invention are all incremental);

8. underdevelopment of innovation rents, permitting rapid copying and extensive multipliers and spillovers, with further impacts on the interaction between the formal and informal economies and the state;

9. exit and exodus as intrinsic to innovation;

10. creative destruction (of child labour, work, technologies, cultures, and relations of exchange, products, forms of organization, and forms of finance); and

11. continuity of the informally acquired knowledge base as vital to the engagement with novelty.

Far removed from the stereotype of 'rural idiocy'[19] and the stagnation and conservatism of 'subaltern towns', these are all evidence of the mobilization of abundant social and economic resources and relations.

Key institutions which regulate innovation in the informal economy, incentivizing or discouraging novel practices, have been found to include the family firm, social identity, education (formal and informal), banks (formal) and informal finance, and more or less registered and active business associations.

Is there an informal structure of innovation resembling an innovation system? The meshing of formal and informal institutions counters the idea that they are discrete epistemological universes (Gupta 1999), but this does not mean that the formal/informal distinction should be abandoned, or that this hybrid 'ecosystem' works in a coordinated or systematic way, or that it is immune from contradictions. Institutions serving useful roles in the structure of accumulation (gender, for

---

[19] For Marx, who coined the phrase, this meant isolation, ignorance, and an inability to cooperate.

instance) may be barriers to agency and innovation. A coherent informal structure of innovation has yet to emerge and may never do so.

References

Basile, E. 2013. *Capitalist Development in India's Informal Economy*. London: Routledge.

Basile, E. and B. Harriss-White. 2000. 'Corporative Capitalism: Civil Society and the Politics of Accumulation in Small Town India', Working Paper No. 38, Queen Elizabeth House Working Paper Series, Oxford University.

Birtchnell, T. 2013. *Indovation: Innovation and a Global Knowledge Economy in India*. Basingstoke: Palgrave Macmillan.

De Soto, H. 2000. *The Mystery of Capital*. New York: Basic Books.

DeLong, J.B. 2003. 'India since Independence: An Analytical Growth Narrative', in D. Rodrik (ed.), *In Search of Prosperity: Analytical Narratives on Economic Growth*, pp. 184–204. Princeton, NJ: Princeton University Press.

Denis, E., P. Mukhopadhyay and M-H Zérah. 2012. 'Subaltern Urbanisation in India', *Economic and Political Weekly*, 47(30): 52–62.

Dickson, D. 1988. *The New Politics of Science*. Chicago: University of Chicago Press.

Flyvbjerg, B. 2006. 'Five Misunderstandings about Case-study Research', *Qualitative Inquiry*, 12(2): 219–45.

Food and Agriculture Organization of the United Nations (FAO). 2012. 'Agricultural Innovation in Family Farming'. Available at http://www.fao.org/nr/research-extension-systems/ais-ff/en/, last accessed on 15 March 2017.

Gathorne-Hardy, A. 2013. 'Greenhouse Gas Emissions from Rice', RGTW Working Paper No. 3. Available at http://www.southasia.ox.ac.uk/working-papers-resources-greenhouse-gases-technology-and-jobs-indias-informal-economy-case-rice, last accessed on 15 March 2017.

Gupta, A. 1999. 'Science, Sustainability and Social Purpose: Barriers to Effective Articulation, Dialogue and Utilization of Formal and Informal Science in Public Policy', *International Journal of Sustainable Development*, 2(3): 368–71.

Harriss, B. 1972. 'Innovation Adaption in Indian Agriculture: The High Yielding Varieties Programme', *Modern Asian Studies*, 6: 71–98.

———. 1981. *Transitional Trade and Rural Development*. New Delhi: Vikas.

———. 1991. 'The Arni Studies: Changes in the Private Sector of a Market Town, 1971–1983', in P. Hazell and C. Ramasamy (eds), *The Green Revolution Reconsidered*, pp. 181–212. Baltimore: Johns Hopkins.

Harriss-White, B. 2003. *India Working: Essays in Economy and Society*. Cambridge: Cambridge University Press.

————. 2011. 'Theoretical Plurality in Markets Conceived as Social and Political Institutions', in J. Gertel (ed.), *Economic Spaces of Pastoral Production and Commodity Systems: Markets and Livelihoods*, pp. 25–42. London: Ashgate.

————. 2012. 'Capitalism and the Common Man', *Agrarian South*, 1(2): 109–60.

————. 2014. 'Labour and Petty Production', *Development and Change*, 45(5): 981–1000.

———— (ed.). 2016. *Middle India and Urban–Rural Development: Four Decades of Change*. New Delhi: Springer.

Harriss-White, B. and G. Rodrigo. 2013. '"Pudumai"—Innovation and Institutional Churning in India's Informal Economy: A Report from the Field', Paper presented at Innovation in India conference, University of Pennsylvania, 13–15 November, available at http://www.southasia.ox.ac.uk/sites/sias/files/documents/PUDUMAI%20-%20INNOVATION%20AND%20INSTITUTIONAL%20CHURNING%20IN%20INDIA%E2%80%99S%20INFORMAL%20ECONOMY%20a%20report%20from%20the%20field.pdf, last accessed on 15 March 2017.

Harriss-White B. and A. Sinha. 2007. 'Introduction', in B. Harriss-White, and A. Sinha (eds), *Trade Liberalization and India's Informal Economy*. New Delhi: Oxford University Press.

International Bank for Reconstruction and Development (IBRD). 2012. *Agricultural Innovation Systems: An Investment Sourcebook*. Washington, DC: World Bank.

Jayaraj D. and K. Nagaraj. 2006. 'Socio-economic Factors underlying the Growth of Silk-weaving in the Arni Region—A Preliminary Study', Monograph No. 5 in S. Guhan Memorial Series, Madras Institute of Development Studies, Chennai.

Kanbur, R., S. Lahiri, and J. Svejnar. 2012. 'Informality, Illegality and Enforcement: Introduction', *Review of Development Economics*, 16(4): 511.

Kline, S.J. and N. Rosenberg. 1986. 'An Overview of Innovation', in R. Landau and N. Rosenberg (eds), *The Positive Sum Strategy: Harnessing Technology for Economic Growth*, pp. 275–305. Washington, DC: National Academies Press.

Larsen, J. 1992. 'Informality, Illegality and Inequality', *Yale Law and Policy Review*, 20(1): 137–82.

Leys, C. and B. Harriss-White. 2012. 'Commodification: The Essence of Our Time', *Open Democracy*, 2 April. Available at https://www.opendemocracy.net/ourkingdom/colin-leys-barbara-harriss-white/commodification-essence-of-our-time, last accessed on 15 March 2017.

Lundvall, B.-A. 1992. *National Systems of Innovation*. London and New York: Pinter.

Messadri, A. 2010. 'Globalisation, Informalisation and the State in the Indian Garment Industry', *International Review of Sociology*, 20(3): 491–511.

Mohan, A.K. and C. Rajagopal. 2010. 'Outsourcing Governance? Revising the Master Plan of Bangalore', Paper presented at the 46th ISOCARP Congress, Nairobi, Kenya. Available at http://www.isocarp.net/Data/case_studies/1810.pdf, last accessed on 15 March 2017.

Olsen, W.K. and J. Morgan. 2010. 'Institutional Change from within the Informal Sector in Indian Rural Labour Relations', *International Review of Sociology*, 20(3): 535–55.

Polzin, C. 2007. 'Credit in South India: An Empirical Critique of Competing Frameworks of Institutional Change', MPhil thesis, Queen Elizabeth House, Oxford University.

Radjou, N., J. Prabhu, and S. Ahuja. 2012. *Jugaad Innovation: A Frugal and Flexible Approach to Innovation for the 21st Century*. New York: Random House.

Rodrik, D. and A. Subramanian. 2004. 'From Hindu Growth to Productivity Surge: The Myth of the Indian Growth Transition', NBER Working Paper No. W10376.

Roman, C. 2008. 'Learning and Innovation in Clusters: Case Studies from the Indian Silk Industry', Doctoral dissertation, Oxford University, Oxford, UK.

Ruthven, O. 2008. 'Metal and Morals in Moradabad: Perspectives on Ethics in the Workplace across a Global Supply Chain', Doctoral dissertation, Oxford University, Oxford, UK.

Sen, K. 2014. 'The Indian Economy in the Post-reform Period: Growth without Structural Transformation', in D. Davin and B. Harriss-White (eds), *China–India: Pathways of Economic and Social Development*, Proceedings of the British Academy 193, pp. 47–62. Oxford: Clarendon Press.

Shah, T. and S. Verma. 2008. 'Co-management of Electricity and Groundwater', *Economic and Political Weekly*, 43(7): 59–66.

Sharma, K. 2009. 'Slumdogs and Small Towns', *Infochange*, April. Available at http://infochangeindia.org/urban-india/cityscapes/slumdogs-and-small towns.html, last accessed on 15 March 2017.

Srinivasan, M.V. 2010. 'Segmentation of Urban Labour Markets in India: A Case Study of Arni, Tamil Nadu', Doctoral dissertation, Jawaharlal Nehru University, New Delhi, India.

Te Lintelo, D. 2009. 'The Spatial Politics of Food Hygiene: Regulating Small-scale Retail in Delhi', *European Journal of Development Research*, 21(1): 63–80.

White, R., A. Gathorne-Hardy, B. Harriss-White, and R. Hema. 2012. 'Resources, Greenhouse Gas Emissions, Technology and Work in Production and Distribution Systems: Materiality in Rice in India', Working Paper No. 1. Available at http://www.southasia.ox.ac.uk/working-papers-resources-greenhouse-gases-technology-and-jobs-indias-informal-economy-case-rice, last accessed on 31 May 2017.

3

TARUN KHANNA
BUDHADITYA GUPTA

The Private Provision of Missing Public Goods

Evidence from Narayana Health in India*

The role of business in society has long been debated (Davis 1973). On the one hand, the Friedman perspective (1962, 1970) argues that governments have the responsibility for social development and business should solely focus on creating wealth for shareholders. On the other hand, scholars have emphasized the need for business to engage in social activities primarily based on principles of morality (Donaldson and Dunfee 1999; Goodpaster 1991; Jones and Wicks 1999) or legitimacy (Post, Preston, and Sachs 2002; Wood 1991). This chapter explores this long-standing dilemma in the context of resource-constrained entrepreneurs in developing countries. The combination of low levels of socio-economic development and resource-constrained entrepreneurs provides us with an opportunity to learn from a situation that, to the mainstream academic literature genres, is an 'extreme' one,

* A conference, India as a Pioneer of Innovation: Constraints and Opportunities, 14 and 15 November 2013, Penn Law, University of Pennsylvania, provided invaluable comments. See https://casi.sas.upenn.edu/events/india-innovation2013, last accessed on 15 March 2017.

though it is quite commonplace in the developing world (Eisenhardt 1989; Pettigrew 1988).

Khanna and Palepu (1997) have developed the idea of institutional voids as the defining characteristic of emerging markets. To understand the utility of this construct in the context of this chapter, consider the following thought experiment (Khanna 2014). Suppose, as a golf aficionado, you take down the fence around a large open field. You still will not have a golf course. For it to be one, you will need flags and holes, closely trimmed greens, meticulously planned fairways, and a clubhouse that creates and enforces a world-class golfing culture. Some of this is easier to do in short order, usually the 'harder' stuff, and some takes time to emerge, usually the 'softer' stuff. It is the same with markets. Simply removing excessive red tape and dismantling barriers helps, but is insufficient because a host of specialist functions are needed to identify would-be buyers and sellers—economists call these 'information problems'—and to build the trust between them so that any transaction they consummate will be honoured. This trust can be enforced in various ways and is often studied under the rubric of 'property rights'.

In more developed settings, specialized intermediaries facilitate information flows between the contracting parties. Besides, capital markets have rating agencies, accountants, regulators, banks, and insurance companies that enable the flow of funds amongst parties, while schools, unions, and unemployment insurance ease the relationships between labour and employers. However, in emerging markets, these specialized intermediaries either do not exist or do not work properly. This situation prevents buyers and sellers from transacting efficiently and inexpensively.

In *Winning in Emerging Markets*, Khanna, Palepu, and Bullock (2010) developed a taxonomy of four types of institutional mechanisms, or specialized intermediaries, which underpin the efficient working of markets. Equivalently, the absence of varying subsets of these is referred to as 'institutional voids', while their presence renders a market an 'emerging' one for definitional purposes. These specialized intermediates are:

1. *Credibility enhancers*: These include accreditation agencies, rating agencies, and auditing firms which lend credibility by independently corroborating sellers' claims.

2. *Information analysers*: The more information available to market participants, the better the decisions they make and the smoother the market functions. In this regard, the role of financial analysts, media rankings, and entities like 'consumer reports' is vital.
3. *Aggregators and distributors*: Big-box retailers, financial institutions, and talent development and placement agencies bring buyers and sellers together in an efficient way.
4. *Transaction facilitators*: Credit card providers, clearing institutions, brokers, and employment exchanges further help to oil the wheels of commerce.

In addition to these four types of market institutions, public sector institutions perform two other essential roles: (a) they regulate markets; and (b) they adjudicate disputes through government regulators, consumer protection agencies, and judicial systems. The extent and efficiency of these intermediary services axiomatically characterize any imperfectly functioning market for products and services, talent, capital, and even ideas.

The institutional inadequacies in emerging markets impede routine business activities. For example, how do businesses find, hire, and maintain a productive workforce in the absence of robust skill development programmes? Similarly, capability gaps in related industries that provide essential inputs constrain a business's ability to create and deliver value to the end customer. How does the entrepreneur raise capital, define the target customer segment, estimate buying ability and collect revenue, attract and train talent, signal quality, and manage numerous other such routine activities?

Cumulatively, such difficulties are often extreme, and it is often not obvious how businesses in developing countries should respond. Should they refrain from actively addressing these gaps and hold the government responsible for developing solutions to the institutional voids? It is indeed common for many businesses in developing markets to 'just manage' with the infrastructure and services provided by the state; typically, the belief in these situations is that low productivity and poor quality are natural corollaries of being situated in underdeveloped regions. At the other extreme, businesses could actively engage in development activities—some of a purely social nature, others with direct economic implications—in an attempt to change the overall business landscape. The choice requires an assessment of the scale of effort and resources necessary for the businesses to make a sustainable impact, given that in many emerging markets, dysfunctional or misguided state intervention complicates the picture.

Khanna's prior work has highlighted one commonly found solution across developing countries, centred on so-called business groups. These are organizational structures comprising a diverse set of businesses, often controlled by a single family, and bound together by common ownership and board membership (Khanna and Palepu 1997, 2000; Khanna and Rivkin 2001; Khanna and Yafeh 2007). Such groups are able to 'fill' institutional voids by developing markets 'internal' to their organizations, to compensate for the absence of well-functioning external markets (indeed, that is part of the raison d'être for the business groups). Fisman and Khanna (2004), for example, show that groups had bottom-line (accounting) benefits to investing in so-called 'social services' in remote, underdeveloped parts of India, since these investment and operating costs were adequately offset by their resultant disproportionate ability to attract talent to move to these areas. Practically, just think of a scarce managerial talent moving to a setting without amenities in the absence of someone going out of the way to provide healthcare and education for her family.

In the case discussed in this chapter, a similar mechanism is underway, even though we are not considering a business group here in the traditional sense. For an entrepreneur to gain traction and scale, investments are needed in seemingly ancillary activities that often seem to be of a more 'social' than economic nature, and, to traditional academic eyes, should purely be the preserve of the provider of public goods, the state. But the state is somnolent or worse, so that is not a viable option. It turns out that these social investments by the private entrepreneur secure both private and public economic benefits.

We argue this point of view based on observations from our multi-year research at Narayana Health (NH), a chain of hospitals in India (Khanna and Bijlani 2011; Khanna and Gupta 2014; Khanna, Rangan, and Manocaran 2005). NH had a core animating idea, that of providing low-cost, high-quality tertiary cardiac care to a large number of poor patients, but, for the animating idea to be realized and to get these patients to the doorsteps of the hospital(s), a number of institutional voids had to be filled. These were done with entrepreneurial aplomb, sometimes even goading the state into action, by being a trusted intermediary to providers of all sorts of factor inputs that would otherwise not make their services available. This amounts to partial private provision of public infrastructure.

Let us further elaborate on the idea of partial private provision of public infrastructure. Consider a classic public good, like a road, or an elementary school. Those who use it benefit from it, but it also benefits myriad others. No individual will ever benefit from a road or a school sufficiently to invest in it as much as society at large could justify investing. So, traditionally, public infrastructure is invested in by the state. But when the state does not act, it is better to have at least some investment by some private entity than no investment—perhaps by one that captures enough of the benefit for it to justify incurring the investment and effort, even if it is not the lion's share. Hence, it is called a 'private provision', and it is considered 'partial' because this private entity still will not invest as much as the state might.

Narayana Health—A Background

Narayana Health was founded in 2001 by Dr Devi Shetty in Bangalore (now Bengaluru), India, for delivering high-quality care at low costs—initially, cardiac care for the poor. On completion of his medical studies in India, Dr Shetty trained in cardiac surgery at Guy's Hospital in the United Kingdom (UK) from 1983 to 1989. He returned to India in 1989 and worked at private hospitals in Kolkata and Bengaluru till 2000.

During his tenure at these private hospitals during the 1990s, he observed that few in India could afford cardiac surgeries. Shetty (2014) recollected:

I would see patients all day and some of them needed cardiac surgeries. But, surprisingly, none of them came back for the surgery. Over time, I realized that they just could not afford the cost of a cardiac surgery. Almost 100 years after the first heart surgery, only 10 percent of the world's population can afford a heart surgery. The rest, if they ever need a cardiac surgery, gradually perish.

Thus, when Dr Shetty started the 150-bed NH in Bengaluru, he knew that innovative mechanisms would be critical to address both the accessibility and affordability gaps burdening the healthcare system in India; these efforts could not be confined within the four walls of the hospital. Effectively, barriers had to be circumvented to allow poor patients seeking cardiac care to come to the hospital. These barriers were diverse and included informational, physical, financial, and emotional obstacles.

Given NH's efforts to address these barriers and the high demand for affordable, quality cardiac care in the region, the cardiac hospital at Bengaluru grew rapidly to house 500 beds, 10 operating theatres, two cardiac catheterization labs, and its own blood and valve banks. The physicians, NH employees earning attractive fixed salaries, worked long hours, while the surgeons did an average of 10–12 cardiac surgeries a week, compared to a weekly average of two to four surgeries by a typical surgeon in the United States (US). The resultant learning led to excellent outcomes: specifically, for coronary artery bypass graft procedures, a common cardiac surgery across the world, NH had a 1.27 per cent mortality rate and 1 per cent infection rate, which was comparable to rates of 1.2 per cent and 1 per cent, respectively, in the US. The scale also led to efficiencies in procurement and utilization of expensive medical devices (for example, CT scan machines). By 2013, the volume of cardiac surgeries at the NH Bengaluru had reached 7,800 (5,000 adult and 2,800 paediatric cases) and 17,600 catheterization lab procedures. These volumes exceeded those in most, if not all, hospitals in the world.

To provide affordable cardiac care to the masses, NH followed a hybrid strategy of attracting paying patients by virtue of its reputation for high quality, and combining this appeal with a relentless focus on lowering its costs of operation wherever possible, to ensure that a larger number of people could seek treatment.[1] The surplus gained from patients who could pay was used to subsidize procedures that were performed at or below cost for patients who could not afford the full fee. The philosophy of the hospital was that no patient should be turned away because of an inability to pay. The hospital offered a

[1] This core NH model has been described in detail in Khanna and Gupta (2014) and Khanna, Rangan, and Manocaran (2005).

scheme called Karuna Hrudaya, which allowed financially constrained patients to pay INR 65,000 (USD 1,400 as per conversion rates of 2004) per open-heart surgery, with NH absorbing the remaining costs (Khanna, Rangan, and Manocaran 2005). For patients who could not afford this package, the Narayana Hrudayalaya Trust, a charitable organization with offices within the hospital, helped to arrange funds from a general corpus or by specifically seeking donations from a list of individuals and organizations.

In the following sections, we review, sequentially, some of NH's key initiatives focused on addressing the institutional barriers preventing poor patients in need of cardiac care from coming to NH; these have also been discussed in Khanna and Gupta (2014) and Khanna, Rangan, and Manocaran (2005). Table 3A.1, presented in Appendix 3A, relates each of these to the particular institutional voids for which it compensates.

Health Insurance

Looking at the plight of the poor farmers in India, Dr Shetty realized the need for a low-cost health insurance programme. The Yeshasvini insurance programme, started in 2002 with the support of the local government, allowed its 4 million members access to more than 1,600 surgical treatment modalities by paying a premium of INR 5 (11 cents at exchange rates at the time) a month. The insurance members were farmers belonging to different cooperatives located in Karnataka; any farmer who was part of a cooperative for at least a year was eligible to participate in the insurance programme.

Dr Shetty was aware that microinsurance schemes in many developing countries were managed unsystematically and thus had resulted in limited success. This motivated him to leverage the existing state-controlled cooperative societies of Karnataka in setting up the insurance programme. The principal secretary of the cooperative department (of the Government of Karnataka), A. Ramaswamy, lent his support to the project and arranged for the involvement of S.M. Krishna, the then chief minister. Yeshasvini was launched as a state programme, with the government contributing INR 2.5 for every INR 5 paid by the farmers. Moreover, in order to keep the upfront collection costs to a minimum, the state government made available the infrastructure of its post offices (operated by the government) to collect the INR 5 premium, track monthly payments, and issue a Yeshasvini member card. The initial task of getting the hospitals to participate and selling the idea to cooperatives was conducted by the trust which NH established to manage Yeshasvini, but, over time, the daily operations were handed over to a third-party administrator, which also coordinated payment to hospitals.

Though Dr Shetty and others at NH put in significant effort to launch Yeshasvini, they were well aware that the involvement of the government was critical in making the effort successful. The farmers targeted by the insurance programme were more likely to trust government agencies instead of a private organization. At the same time, the reputation gained by Dr Shetty and NH while delivering high-quality care to patients in Kolkata and Bengaluru, specially to the poor, also helped get support from the farmers, the hospitals, and the state government. That Dr Shetty's reputation as a surgeon and socially responsive entrepreneur facilitated efforts in establishing mass insurance schemes is no surprise, given that reputation has been considered as a key building block of business relations and performance (Carmeli and Tishler 2004; Dierickx and Cool 1989; Fombrun 1996; Fombrun and Shanley 1990; Hall 1993; Rhee and Haunschild 2006; Roberts and Dowling 2002), and especially given that reputation forms the source of sustained advantage in emerging markets due to the absence of other formal credibility-enhancing intermediaries.

The Yeshasvini programme allowed thousands of farmer families to avail themselves of tertiary care free of cost at different participating hospitals. Dr Shetty (2014) remarked, 'Poor people on their own are weak, but when they come together they can solve many problems.' The programme also allowed NH and other hospitals in the region to benefit from increased patient flow. Over time, as Yeshasvini grew in members, the insurance programme was taken over and managed by the state of Karnataka. Moreover, its success and popularity resulted in Karnataka and other Indian states coming up with similar insurance programmes for different segments of the indigent population, such as expecting mothers, children, and senior citizens.

Of course, the insurance programme is only a small part of a constellation of efforts that made care financially available. As described earlier, a larger portion was the cross-subsidy model that NH developed and that has persisted over the past decade, whereby patients who can afford to pay more than the break-even surgery cost do so and the resulting surplus, along with additional charitable donations, is deployed to care for the indigent.

Medical Training

Dr Shetty realized that he had to combat the incredible scarcity of cardiologists and other physicians to improve access in the long term. From the beginning, the doctors at NH were actively involved in training the next generation of specialists. NH ran 19 postgraduate programmes for doctors and other medical staff, including diplomas in cardiac thoracic surgery, cardiology, and medical lab technology. The hospital also offered the country's only formal training programme for paediatric cardiac surgery, reflecting the rich expertise of NH's doctors in paediatric care. NH also had a two-year training programme for Bachelor of Medicine and Bachelor of Surgery (MBBS)-qualified doctors (the Indian equivalent to a Doctor of Medicine [MD]). The general practitioners (GPs), who were trained in this programme, had the knowledge and skills to manage emergency and non-interventional cardiology cases, and this increased the general population's access to quality cardiac care.

Beyond training doctors, a separate department at the hospital coordinated the training of nurses. The quality of nurses available in the market varied significantly given the relatively weak regulatory and clinical training standards and compliance. From the beginning, NH emphasized the critical role of the entire clinical team in delivering successful outcomes. All nurses at NH were required to complete a year of training, which included a minimum six-month period in a critical care unit. In addition to training nurses specialized in cardiac care, the nursing college housed within NH offered degree and diploma courses for trainee nurses. To encourage students from poor remote areas, who would benefit most from these educational opportunities, NH arranged for nursing student loans from government-controlled banks to cover fees and living expenses. During the course, the trainee nurses assisted senior nurses at the hospital and, on graduation, worked at the hospital for up to two years, thus supplying much-needed skilled manpower to NH.

In the absence of adequate aggregators like medical training schools that ensure the availability of trained clinical staff, NH set up its own training programmes. Moreover, by providing financial support and delivering high-quality training in an environment where such opportunities were limited, NH ensured a steady inflow of motivated resources interested in developing clinical skills.

Telemedicine

As in other developing countries, the population in rural areas did not have access to quality healthcare due to lack of trained physicians in these regions. Since cardiac specialists were rare in remote areas, heart attack victims usually turned to GPs, who sometimes prescribed incorrect treatment due to the lack of knowledge or facilities to correctly diagnose the problem. Sensing the need for immediate treatment and care in rural areas, Dr Shetty set up nine coronary care units (CCUs) across India, linked to NH. Each CCU was equipped with beds, medication, computers, electrocardiogram (ECG) machines, video conferencing devices, and technical staff trained to operate the equipment. In addition, NH trained the GPs at the CCUs to perform checks on patients and to administer treatment. With help from S.N. Informatics, a software development company located in Bengaluru, NH also created a software programme that allowed ECG images to be scanned and transmitted via a Web connection.

When a patient visited a CCU, the GP on duty first took an ECG, which was transmitted to a specialist at NH. With the patient and the GP on the screen, the specialist then diagnosed the condition and advised the GP on the appropriate treatment. Patients who needed to be kept under observation stayed overnight at the CCUs and specialists at NH conducted daily virtual checks on their condition. In serious cases, the patient, once stabilized, was transferred to NH for surgery. In very remote areas where video conferencing facilities could not be set up, a network of around 100 family physicians were still able to use the software to transmit ECG images for diagnosis at NH.

Although telemedicine was not a new concept in India in 2001 (the Apollo Group of hospitals started using telemedicine earlier), NH hospitals made up the country's largest network and were the only hospitals in India that provided the service for free. The cost of setting up the CCUs was funded by Asia Heart Foundation, as were the staff salaries and operational costs. Almost from the beginning, the project was supported by the Indian Space Research Organisation (ISRO), a government agency which adopted telemedicine as a community project and provided connectivity for the CCUs free of charge. ISRO's technology allowed telemedicine to operate by satellite connection, thus providing clearer images than the more expensive (and less reliable) phone lines—'the patient must see the compassion on the doctor's face', said Dr Shetty (Khanna, Rangan, and Manocaran 2005).

In addition, the Government of Karnataka was so enthusiastic about NH's work that the state planned to sponsor 29 additional CCUs. Between 2001 and July 2004, the NH facility performed 9,591 tele-consultations and the CCUs had 4,077 in-patients, many of whom would not have received treatment otherwise. The telemedicine units were also linked to a clinic in Malaysia, a children's cardiac facility in Mauritius, and a medical school in Hanover, Germany, with plans for new connections with Bangladesh, Tanzania, and Pakistan. Also, Dr Shetty believed the concept could be extended across other medical areas. 'If the patient does not require surgery, then the doctors may not need to touch him ... once thousands of CCUs are networked, telemedicine can be self-sustaining for a few rupees per patient' (Khanna, Rangan, and Manocaran 2005).

Transportation

In addition to the hospital in Bengaluru, Dr Shetty also started a heart hospital in Kolkata in 2001. Both the hospitals organized outreach camps for cardiac diagnosis and care. Each weekend, two buses were sent out to rural areas, up to 800 km away from the hospitals in Bengaluru and Kolkata. Each bus was staffed with at least three doctors, including an experienced cardiologist, and two technicians trained to perform echocardiograms. In order to ensure that the best possible diagnosis could be made on site, the buses were outfitted with echocardiography equipment, a treadmill, a defibrillator, ECG machines, equipment needed for resuscitation in emergencies, and a generator to ensure the machines could be used in rural areas with irregular power supply.

The outreach camps were organized by local associations (for example, the Lions Club or Rotary Club) who advertised the day-long event in advance and arranged for patients to attend. On average, each camp screened 400 people a day, none of whom were required to pay either the hospital or the organizers. When a diagnosis indicated a need for medical intervention, the patient was advised to visit NH's hospitals in Bengaluru or Kolkata, where the procedure was performed at or below the cost price with help from the NH charitable trust.

Social Services

NH's initial effort in establishing the heart hospital and providing care to the indigent population through the cross-subsidy model catalysed other entrepreneurs. As patient volume grew at the Bengaluru hospital, small businesses grew up around the hospital to provide lodging, food, and transportation to patients and their families visiting from all parts of India and abroad. The reputation and success of NH ensured guaranteed customer volume for these small businesses and defrayed the usual business risks. In essence, the NH model provided the platform that allowed other entrepreneurs to set up successful businesses catering to the socio-economically disadvantaged customer segment in India.

Other than enabling a supporting environment for other start-up businesses, NH provided opportunities to a number of private enterprises and wealthy individuals to engage in social service. Over time, as NH's reputation grew, these enterprises and individuals, often themselves beneficiaries of high-quality healthcare at Dr Shetty's hospital, started contributing to cover treatment costs of economically disadvantaged patients. This provided an alternative third mode of finances for healthcare for the indigent population and complemented the state-managed insurance schemes and NH's cross-subsidy model.

Finally, NH itself got involved in other social activities. Dr Shetty had always believed that if children from poor families became doctors, they would be willing to serve the rural poor, which would in turn improve the accessibility and affordability of healthcare for a significant population base. With that objective, NH mentored and offered scholarships to select school-age village children once they committed to pursuing medicine as a career.

Incubator

Medical technology and devices are the critical backbone for healthcare delivery today. However, most of the technology and devices are developed in the US and Europe, where the focus is typically on improving efficacy, with little consideration to cost. This results in med-tech offerings that do not fit well in the context of resource-constrained emerging markets. From the beginning, Dr Shetty realized the importance of partnering with local entrepreneurs and organizations in developing products that support the delivery of affordable care. Numerous start-ups established temporary offices in the basement of NH and partnered with NH in an effort to develop products appropriate for the Indian market. The high-volume clinical setting and the rich experience of NH physicians provided an ideal environment for product development by these start-ups. Over the years, these products have varied from affordable disposable surgical scrubs to more complex technology solutions for clinical treatment of particular diseases and software solutions to manage patient care in the critical care units in hospitals. In some of these cases, NH would adopt the products for use at the hospital, and also provide support to the entrepreneurs in their commercialization efforts.

Overall, in its role as an incubator, NH helped entrepreneurs manage critical questions related to capital, talent, property rights enforcement, quality accreditation, key client access, and more, and thus addressed the numerous institutional voids that commonly challenge start-up organizations in developing economies like India.

The NH Group Today

By 2011, the cardiac hospital at Bengaluru had evolved into a multi-specialty 'health city' that consisted of a 25-acre campus and housed a 900-bed heart hospital, a 1,400-bed cancer hospital, a 500-bed ortho-paedic and trauma hospital, a 300-bed eye hospital, an organ transplant institute, and departments in neurosurgery, neurology, paediatrics, nephrology, urology, gynaecology, and gastroenterology. NH also rep-licated the multi-specialty health city model in other cities like Kolkata and Ahmedabad and began to build 200- to 300-bed general hospitals in smaller Indian cities. The clinical outcomes realized at different hospitals, across different tertiary specialties, were comparable to the leading hospitals in developed countries. In January 2011, the cardiac hospital at Bengaluru was accredited by the US-based Joint Commission International (JCI), the not-for-profit global division of the Joint Commission Resources that was established in 1994 to help healthcare organizations around the world to improve their performance. The accreditation by JCI validated the high-quality care at NH, which led to an increased number of patients visiting NH from countries in South Asia, the Middle East, and Africa.

By 2014, NH was managing 7,000 beds in 27 hospitals at 16 different cities in India; this made it the third-largest hospital group in India. The scale of its operations and reputation allowed it to serve patients from different socio-economic classes and at the same time, develop deep expertise in different super specialties like oncology and interventional radiology. Moreover, the same year, NH opened a 104-bed hospital in the Cayman Islands focused on cardiac care and orthopaedics; the long-term intent was to develop the Cayman facility into a 2,000-bed health city over the next few years and serve patients from the Caribbean, North America, and South America.[2]

***

Narayana Health's attempt to improve the affordability and accessibility of healthcare went beyond managing processes and costs within the four walls of the hospital.

The effort of setting up of health insurance schemes, training man-power, subsidizing costs for poor patients, hosting camps at remote

---

[2] Further details of NH's internationalization are discussed in Khanna and Gupta (2014).

locations, providing telemedicine support, and incubating development of appropriate technology illustrates how infrastructure development activities by start-ups in emerging markets often align with core business goals. Figure 3.1 illustrates NH's attempts to compensate for the institutional voids that would otherwise impede its vision of providing scalable heart surgery to the mass indigent. Each of these initiatives can arguably be seen to be the preserve of the state. However, the entrepreneur, Devi Shetty, was faced in each instance with either waiting (indefinitely?) for the state or incurring investments to compensate for the institutional void in question. Sometimes the investments were made in tandem with state organizations, effectively mobilizing the somnolent agency in question.

Of course, those efforts, whether 'solo' or in tandem with the state, have spillover benefits to society in general. For example, doctors and nurses trained by NH often left the organization to work in the wider world; the insurance programme provided a template that many other organizations and state governments imitated; and so on. But it is also the case, in at least several of these instances, that enough private benefits accrued to the NH team to make it even economically worth their while. As in Fisman and Khanna (2004), this partial private provision of public infrastructure by NH illustrates an interesting manner by which public goods are provided in developing countries with poorly functioning states.

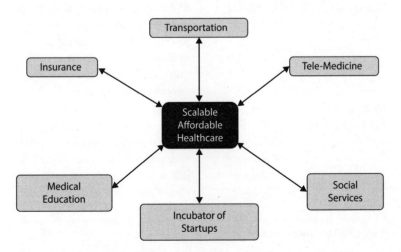

FIGURE 3.1   NH Social Initiatives
Source: Created by authors.

Appendix 3A

TABLE 3A.1   Institutional Voids Addressed by NH Initiatives

NH Initiative	Roles	Effect
Health Insurance	• Aggregator • Credibility Enhancer	Aggregated the buying power of the poor so as to afford quality health-care; Yeshasvini was a trailblazer in medical insurance industry for the poor.
Transportation	• Aggregator	Aggregated disparate patient pools from the hinterland to bring them to the hospital in a cost-effective way.
Telemedicine	• Information Analyser	Sorted out which patients needed to come to the main hospital and which could be treated differently or elsewhere.
Social Services	• Transaction Facilitator	NH became a transaction platform, matching indigent patients in need of social services with charities and other entities that sought to provide these services.
Medical Education	• Aggregator • Credibility Enhancer	Allowed the development of clinical staff in a high-volume clinical setting; provided financial support; and access to a high volume of patients led to improved clinical expertise and, over time, the NH training became associated with high quality.
Medical Incubator	• Aggregator Information Analyser • Credibility Enhancer	Med-tech start-ups thrive due to easy access to clinical know-how and informal mechanisms and competitions promoted to attract talent and risk capital; NH gets access to fresh ideas, filling a void in the risk capital market.

Source: Authors' own research.

References

Carmeli, A. and A. Tishler. 2004. 'The Relationships between Intangible Organizational Elements and Organizational Performance', *Strategic Management Journal*, 25(13): 1257–78.

Davis, K. 1973. 'The Case for and against Business Assumption of Social Responsibilities', *Academy of Management Review*, 16(2): 312–22.

Dierickx, I. and K. Cool. 1989. 'Asset Stock Accumulation and Sustainability of Competitive Advantage', *Management Science*, 35(12): 1504–11.

Donaldson, T. and T.W. Dunfee. 1999. *Ties that Bind: A Social Contracts Approach to Business Ethics*. Cambridge, MA: Harvard Business Press.

Eisenhardt, K.M. 1989. 'Building Theories from Case Study Research', *Academy of Management Review*, 14(4): 532–50.

Fisman, R. and T. Khanna. 2004. 'Facilitating Development: The Role of Business Groups', *World Development*, 32(4): 609–28.

Friedman, M. 1962. *Capitalism and Freedom*. Chicago: University of Chicago Press.

———. 1970. 'The Social Responsibility of Business is to Increase its Profits', *The New York Times Magazine*, 30 September, pp. 122–6.

Fombrun, C. 1996. *Reputation: Realizing Value from the Corporate Image*. Cambridge, MA: Harvard Business School Press.

Fombrun, C. and M. Shanley. 1990. 'What's in a Name? Reputation Building and Corporate Strategy', *Academy of Management Review*, 33(2): 233–58.

Goodpaster, K.E. 1991. 'Business Ethics and Stakeholder Analysis', *Business Ethics Quarterly*, 1(1): 53–73.

Hall, R. 1993. 'A Framework Linking Intangible Resources and Capabilities to Sustainable Competitive Advantage', *Strategic Management Journal*, 14(8): 607–18.

Jones, T.M. and A.C. Wicks. 1999. 'Convergent Stakeholder Theory', *Academy of Management Review*, 24(2): 206–22.

Khanna, T. 2014. 'Entrepreneurship in Emerging Markets: Contextual Intelligence for the Study of Two-thirds of the World's Population', *Multidisciplinary Insights from New AIB Fellows: Research in Global Strategic Management, Vol. 16*, pp. 221–38. Bingley, UK: Emerald Group.

Khanna, T. and T. Bijlani. 2011. *Narayana Hrudayalaya Heart Hospital: Cardiac Care for the Poor (B)*, Case 712-402. Cambridge, Massachusetts: Harvard Business School Supplement.

Khanna, T. and B. Gupta. 2014. *Health City Cayman Islands*, Case 714-510. Cambridge, Massachusetts: Harvard Business School.

Khanna, T. and K.G. Palepu. 1997. 'Why Focussed Strategies may be Wrong for Emerging Markets', *Harvard Business Review*, 75(4): 41–51.

———. 2000. 'The Future of Business Groups in Emerging Markets: Long-run Evidence from Chile', *Academy of Management Review*, 43(3): 268–85.

Khanna, T., K.G. Palepu, and R. Bullock. 2010. *Winning in Emerging Markets: A Road Map for Strategy and Execution*. Cambridge, Massachusetts: Harvard Business Press.

Khanna, T., V.K. Rangan, and M. Manocaran. 2005. *Narayana Hrudayalaya Heart Hospital: Cardiac Care for the Poor (A)*, Case 505-078. Cambridge, Massachusetts: Harvard Business School.

Khanna, T. and J.W. Rivkin. 2001. 'Estimating the Performance Effects of Business Groups in Emerging Markets', *Strategic Management Journal*, 22(1): 45–74.

Khanna, T. and Y. Yafeh. 2007. 'Business Groups in Emerging Markets: Paragons or Parasites?' *Journal of Economic Literature*, 45(2): 331–72.

Pettigrew, A. 1988. 'Longitudinal Field Research on Change: Theory and Practice', Paper presented at the National Science Foundation Conference on Longitudinal Research Methods in Organizations, Austin, Texas.

Post, J.E., L.E. Preston, and S. Sachs. 2002. *Redefining the Corporation: Stakeholder Management and Organizational Wealth*. Stanford, California: Stanford University Press.

Rhee, M., and P.R. Haunschild. 2006. 'The Liability of Good Reputation: A Study of Product Recalls in the US Automobile Industry', *Organization Science*, 17 (1): 101–17.

Roberts, P.W. and G.R. Dowling. 2002. 'Corporate Reputation and Sustained Superior Financial Performance', *Strategic Management Journal*, 23(12): 1077–93.

Shetty, Devi. 2014. Speech at Cayman Captive Forum 2014, 2–5 December, Cayman Islands.

Wood, D.J. 1991. 'Corporate Social Performance Revisited', *Academy of Management Review*, 16(4): 691–718.

4

PRASHANT KALE
HARBIR SINGH

Innovation in Indian Business Groups

Innovation has been an important agenda item for senior management teams in companies around the world, including those from India. Since the year 2000, India has begun to feature prominently in the many managerial and academic discussions and studies on innovation—this is especially true with respect to the creation of new business models to address the markets at the 'bottom of the pyramid' in emerging economies like India. Other forms of innovation, termed 'reverse innovation' or 'jugaad' (Radjou, Prabhu, and Ahuja 2012), have also been studied and explored. The present chapter complements this body of work by examining innovation in large business groups (BGs), which are an important feature of the business landscape in emerging economies like India (Khanna and Palepu 2000).

The chapter first explains what exactly we mean by the term BGs, why they exist in emerging economies, and some of the benefits and risks for companies affiliated with them. Then it briefly reviews some recent academic work that examines their role and value in the context of innovation. While prior work on BGs provides some interesting insights, it is remarkably silent on the specific organizational mechanisms that BGs can potentially use to enhance and foster innovation. The chapter addresses this gap by providing insights based on fieldwork with a wide range of BGs in India. It explains the different mechanisms they use and how and why they are useful in enabling innovation. It then discusses the challenges that Indian BGs face in this respect and shares some thoughts on future research. While this chapter specifically looks at the role of Indian BGs in innovation, some of the practices and mechanisms it highlights should carry potential trans-boundary relevance as well.

What Are BGs and What Are the Implications of Association with Them?

Diversified BGs play an important role in the socio-economic landscape of many emerging economies (Yiu et al. 2007). Such groups control and coordinate two or more legally distinct firms through commonly held ownership stakes, often complemented by social ties (Guillen 2000; Khanna and Palepu 2000; Morck 2009). Various studies show that these groups are a big part of the business landscape in many countries, especially in emerging economies.

The focus on the growing stream of research on BGs is in order to understand the reasons for their existence, especially in emerging economies, and the benefits or costs for firms affiliated with them (Guillen 2000; Mahmood and Mitchell 2004), as well as the variation in the performance effects of BG affiliation as the quality of the institutional environments in which they are located changes (Chittoor, Kale, and Puranam 2015). Affiliation with a BG potentially provides many benefits to individual firms which are part of such a group. Essentially, it is widely recognized that in the absence of well-developed economic institutions such as capital, labour, and product markets, affiliation may provide firms with access to these resources at low transaction costs (Khanna and Palepu 2000). Moreover, the internal hierarchical control of the BG may also discipline the management of the affiliated firm in terms of how the latter actually utilizes the capital. The BGs may also have superior access to the political power structure in the economy through their consolidated lobbying and influence efforts, and hence benefit from a richer pool of opportunities in the country (Khanna and Rivkin 2001). As with other forms of multi-business organization, BGs can also leverage economies of scale and scope, particularly those of a non-rivalrous nature (Chang and Hong 2000). In addition, BGs and their affiliates also represent a social structure characterized by repeated interaction, family ties, and rich information flows. Consequently, the costs of transacting within the BG may be lower than those for comparable transactions between independent firms (Guillen 2000).

Affiliation with BGs might also entail several costs. The most important of these arise from conflicts between the interests of the BG promoter and those of other shareholders (Keister 1998). Often, BGs have controlling stakes in several firms, but may not have proportionately significant cash flow rights in many of them. This creates incentives for the BG to expropriate and transfer profits from firms in which the BG has low cash flow rights to firms where it has high cash flow rights— a practice known as 'tunnelling' (Morck 2009). Bertrand, Mehta, and Mullainathan (2002) find some evidence of such activity amongst BGs in several countries. The BGs also serve to reduce the bankruptcy and survival risks of weaker affiliates (Chacar and Vissa 2005), but do so at a cost to stronger members, through either cross-subsidization or a general weakening of incentives for managers to run their firms efficiently (Khanna and Rivkin 2001). Like other forms of multi-business organizations, BGs may also be prone to some adverse effects of centralized decision-making that arise in the context of decentralized information (Williamson 1985). For example, group-level initiatives such as poorly conceived corporate training or synergy programmes may impose overhead on all the affiliates without producing offsetting benefits (Khanna and Rivkin 2001). Further, when BGs are essentially family-controlled entities, corporate governance may be subverted through managerial entrenchment. Additionally, controlling families may interfere in the tactical and strategic decision-making of member firms and there is a possibility of nepotism.

Although there is a fair degree of convergence in the literature on the potential benefits and costs of affiliation with BGs, agreement on their net impact on an affiliated firm's performance remains mixed: there is wide variation in the estimated affiliation effects observed across countries and studies (Carney et al. 2011). We argue that, rather than get caught up in the debate of whether BGs are 'paragons or parasites' (Khanna and Yafeh 2007), it is more meaningful to accept that the benefits of BG affiliation may outweigh the costs under certain conditions. Generally, if the primary benefit of affiliation with a BG in emerging economies (and indeed the rationale for their existence) stems from the absence of strong economic institutions, being affiliated with BGs would carry a net positive impact. Additionally, a study (Chittoor, Kale, and Puranam 2015) shows that even if some aspects of the external institutional environment strengthen, as is the case of the capital markets in India, affiliation with BGs is beneficial for firms that are listed in the external capital markets and face the scrutiny of external stakeholders.

Innovation and the Role of BGs

For innovation to happen, a country needs access to some 'innovation infrastructure', that is, a set of resources, such as finance, talent, and technology, in order to undertake innovative activities (Mahmood and Mitchell 2004). While this infrastructure is developed through market-based institutions in developed countries, many emerging economies, in the absence of well-functioning market institutions, are unable to create and provide these aspects. Under such circumstances, BGs step up to address voids, in line with the intermediation role described earlier. Due to underdeveloped capital markets, firms (especially small or new ones) face difficulty in sharing their ideas with potential investors and raising capital to pursue them. In such circumstances, BGs, with their internal capital markets, can act as de facto venture capitalists, allocating capital to new innovative opportunities more effectively and efficiently than the available external markets. Where external capital is available to some extent, BGs are able to use it to support innovation initiatives far more easily than other firms, due to their lower bankruptcy risks. Apart from capital, to successfully pursue innovation, firms also require a pool of innovative or scientific talent. In the face of underdeveloped labour markets in emerging economies, such talent (or access to it) is quite scarce. Here again, BGs provide value by acting as incubators for such a talent pool and incurring the high fixed costs that might be entailed in this endeavour. Because large BGs offer desirable facilities and infrastructure, innovators and scientific talent are willing to accept positions and even intra-group relocations through the internal talent markets that these BGs provide.

When it comes to technology for innovation, firms develop it in-house as well as source it through external partners. Firms in emerging economies have traditionally faced many difficulties with respect to sourcing technology through external partners. Due to the weak intellectual property (IP) regimes and enforcement in most emerging economies, foreign firms have been very hesitant to license their technologies to firms in these countries. But BGs seem to have a small edge here too—foreign firms seem more open to share their technology with BGs (or firms affiliated to them) because, by putting the group's entire reputation at stake, BGs are able to alleviate concerns about IP theft and violation amongst their potential technology partners from overseas. Data on the Tata Group in India as well as BGs in Argentina, Spain, and South Korea show that BGs have been able to form far many joint ventures and alliances than non-BG-affiliated firms to get technology from foreign firms (Khanna and Palepu 2000; Mahmood and Mitchell 2004). Innovation is also more likely when firms are able to recombine knowledge and resources from diverse businesses and sources in novel ways (Grant 1996). By virtue of their presence in multiple business areas along with internal markets and processes to facilitate the sharing and recombination of knowledge from these diverse sources, BGs are better positioned to develop new innovations compared to stand-alone firms that lack these aspects.

Some scholars suggest that, while BGs may foster innovation within firms in their group by providing an innovation infrastructure, they might hamper potential innovation at a macro level. The same factors that enable BGs to create an advantage in the presence of market imperfections, also allow them to create 'entry barriers' for potential (smaller) rivals through preemptive price-cutting in their local businesses (Berger and Ofek 1995). Also, by forming preferential, reciprocal relationships (as buyers or suppliers) with other firms in the group, they potentially foreclose the market to other independent competitors. Collectively, this tendency to create entry barriers can deter independent firms and thus adversely affect innovation in the country, since the diversity of ideas and the competition that new players can bring enhances innovation in the whole economy, and their absence has the opposite effect.

Apart from deterring competition (and hence innovation) at the macro level, BGs can sometimes also deter innovation within their own group of firms. First, the bureaucracy and inertia that sometimes set in within large BGs can work against innovation. Second, the internal capital markets within BGs might also sometimes be prone to agency problems, which, in turn, can lead to less risk-taking and lesser commitment to long-gestation projects, both of which are often true in the case of innovation initiatives. At times, managerial nepotism may also deter talented professionals, which can consequently deter innovation. However, some of the empirical work suggests that, overall, BGs in most emerging markets are more likely to foster innovation than hinder it.

That said, however, extant research is quite silent on the specific internal mechanisms that BGs use to enable innovation in and across group companies. We attempt to address that gap in the literature by examining some of the specific internal mechanisms that can play a positive role in this regard. We do that largely through our fieldwork and interviews of BGs in India.

Indian BGs: A Multipronged Approach to Innovation

Business groups have been a dominant feature of the business landscape in India for more than a century. Table 4.1 provides an overview of the top 10 Indian BGs in terms of revenues and market capitalization. These 10 groups alone account for almost one-sixth of the country's gross domestic product (GDP). While names like Tata, Birla, Bajaj, and Mahindra have long been in this elite list, others, like Vedanta, Essar, and Jindal, are more recent entrants. In the first five decades after India's Independence, most Indian companies, including those affiliated with BGs, focused on either imitating the technology or other innovations of companies from developed economies or 'improvising' on it in creative but incremental ways to compete in India's closed, regulated, and constrained business environment. Only in the last 10–15 years have Indian companies seemed to be moving from improvisation to innovation, to not only meet the growing demands and needs of the Indian customer but also to meet the challenges posed by the foreign and multinational companies entering India in droves to exploit the huge potential of its

TABLE 4.1   Top 10 Business Groups in India as of 2012

Rank	Sales	Market Cap
1	Reliance Industries	Tata Group
2	Tata Group	Reliance Industries
3	Aditya Birla Group	Aditya Group
4	Essar Group	Vedanta
5	O.P. Jindal Group	Mahindra & Mahindra
6	Mahindra & Mahindra	Bajaj Group
7	Bharti Group	O.P. Jindal
8	Reliance ADAG	Reliance ADAG
9	Bajaj Group	Bharti
10	Vedanta	Adani

Notes: Sales / Revenue:
    Cumulative Sales ~ $260 Billion
    Range: $5 Billion - $60+ Billion
    Market Capitalization:
    Cumulative Sales ~ $280 Billion
    Range: $10 Billion - $80+ Billion
Source: India *Business Today*, Nov. 2012, Volume 21, No. 21.

vast but largely untapped domestic market. In this section, we provide an overview of the different 'levers' that Indian BGs have pulled to jump-start their innovation efforts and endeavour to foster innovation on a more continual basis.

Making Innovation Investments

Indian BGs have substantively stepped up their investments in research and development (R&D) and technology development, from just over USD 1 billion in 2003 to over USD 7 billion in 2012. These investments are focused on in-house R&D as well as in-sourcing through technology and licensing agreements (see Figure 4.1). Indian BGs have adopted different organizational approaches to channelling this effort. One set of BGs has focused on setting up group- or corporate-level R&D labs that undertake research in the basic sciences and technologies that are core,

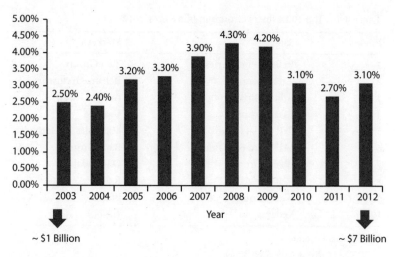

FIGURE 4.1   Indian Business Groups—Growing Investments in R&D and
Technology
Source: Centre for Monitoring of the Indian Economy (2013).

but also common, across many of the different businesses that span the
group. Examples include Mahindra Research Valley of the Mahindra
Group (an approximately USD 250 million investment), the Aditya Birla
Science and Technology Company Private Limited of the Aditya Birla
Group (approximately USD 100 million), and the Innovation Labs set
up by Tata Consultancy Services (TCS) (approximately USD 500 million), which belongs to the Tata Group. As the Mahindra Group has a
large presence in a host of transportation and automotive sectors, its
corporate-level research labs focus on innovation and technology development in core areas relevant to them. The Birla Group, on the other
hand, is present in a range of businesses (cement, aluminium, etc.) that
are very power or water intensive; its science and technology centre is
focused on research and innovation in these areas. This approach is similar to the corporate-level industrial research organizations established
by large United States (US) corporations such as 3M (whose corporate
labs focus on the core science of adhesives) or Corning (whose corporate research is centred around glass materials and technology).

Another set of BGs has taken a different approach to making innovation investments, which is modelled on the 'corporate venturing'

approach widely seen in many companies of the US/developed countries such as Intel, Dell, and Google. For example, the Godrej Group in India, which has a strong presence in a range of agriculture or food businesses, has helped set up a USD 50 million fund called Omnivore Partners that invests in early-stage agri-tech and food technology companies that are innovating new products or business models in areas related to its agriculture and food businesses. The intention here is to give an 'early look' to promising new ideas that can be nurtured more freely outside of its large core businesses but can also be linked to or brought into its own businesses to leverage that innovation at the appropriate stage. The Piramal Science Foundation (USD 30 million) or Shriram Investments (USD 250 million) are some other examples of Indian BGs undertaking this approach. Both approaches play an intermediary role in providing patient and ready capital for innovation when external institutions for doing so are not sufficiently developed in the broader economy.

Establishing Innovation Structures

Innovation investments are not sufficient unless organizations also have a structure to enable idea generation, nurturing, and management in a systematic way (O'Reilly and Tushman 2004). Indian BGs have pursued diverse alternatives and approaches in this regard. Innovative insights often lie at the interface of different knowledge or business streams (Kogut and Zander 1992). Therefore, creating mechanisms to bring together people from diverse areas can certainly help. The Tata Group has set up a Tata Group Innovation Forum (TGIF) comprising senior executives from diverse companies as well as from key corporate functions. This group serves as a forum to generate new ideas based on the amalgamation of insights from different sectors or businesses. It also acts as an effective means to diffuse innovative ideas that might arise in one group company to others through the members of that forum. Thus, the structure facilitates both new idea generation as well as diffusion. Reliance Group's Innovation Council plays a similar role, to some extent.

Innovation initiatives at the group level also need to be championed and guided, and Indian BGs have pursued interesting practices in this direction. India's Godrej Group has appointed one of its most senior executives as the champion and 'point person' for leading group-wide innovation efforts. Interestingly, this person is also the head of human resources (HR) for the entire group. Having the same person oversee both HR and innovation is based on the premise that innovation arises first in the minds of individuals and people, and an executive who oversees both will have a vantage point to create organizational and mentoring structures conducive to helping individual employees and teams become more creative and innovative.

The Tata Group, on the other hand, places the responsibility of enabling innovation in a focused manner on one of its most senior group-level executives, typically a member of the Group Executive Office (GEO) or Group Executive Council (GEC). In 2014, it also appointed a group-level chief technology officer (CTO), the former head of General Electric's (GE's) India-based global research centre, to help drive technology policy and development and accelerate innovation efforts in a more systematic manner. Mahindra Group has gone one step further, creating a structure to handhold group-affiliated companies and teams to understand and learn a set of 'processes and practices' to ensure that all innovation initiatives follow a well-defined process to increase their odds of success. Its dedicated innovation management cell comprises executives who have worked in large, established innovation champions like GE and 3M, to create a formal process and set of tools that Mahindra innovation teams can follow to help them in their endeavour.

Creating an Innovation Culture

Financial investments in R&D and innovation and creation of relevant structures to enable and manage the innovation are mainly the 'limbs' to enable innovation in BGs, but having a culture and climate to foster innovation is the energy for those limbs to function to their full potential. Organization culture refers to the shared values (emphasizing what is important), norms (acceptable attitudes and behaviours), and assumptions that set the context of action. In the transition from improvisation to innovation, Indian BGs have sought to develop a culture that is conducive for innovation. In 2009, the Mahindra Group crafted a new set of values under its 'Rise' initiative, urging employees at all levels and in all companies to emphasize the following beliefs in any action they took:

'accept no limits', 'alternative thinking', and 'drive positive change'.[1] Going beyond simply articulating these aspects, all business units and managers were encouraged to incorporate them into their operating and strategic plans as a way to create a bold and innovative climate. To encourage lateral and diverse thinking, which is the foundation for innovation at the individual level, the company actively added recruits with background and experience in arts and humanities to complement the science, math, and engineering talent of most Mahindra employees. The Godrej Group, which has combined HR and innovation functions under one umbrella, took steps in a similar direction.

Many BGs also established group-wide innovation awards to create buzz and generate a favourable innovation climate. The Tata Group was a leader in this, through its InnoVista annual awards. This programme, which begun in 2006, attracted just a few entries in the first year. By 2012, more than 70 Tata companies were participating, turning in over 2,800 entries for innovative ideas and projects that had been undertaken across the group companies (see Figure 4.2). By capturing innovations

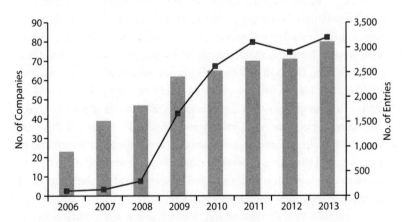

Figure 4.2 Innovation Culture—Tata InnoVista Awards to Celebrate Innovation

Source: Media Releases from www.tata.com.

---

[1] See www.mahindra.com/Rise, last accessed in November 2013.

of Tata companies, recognizing innovators for their enterprise, and encouraging companies to share lessons with each other, the hope was to create a culture of risk-taking amongst group employees and to encourage innovation. The InnoVista awards not only honoured promising innovations that had achieved some degree of success but also had a 'Dare to Dream' set of rewards for sincere and audacious attempts to pursue major innovations that failed to get the desired results. The idea was to underscore the acceptance of 'failure' on the path to innovation and thereby encourage others to do the same and learn from it.

Developing Innovation Capability

Indian BGs also seem to have realized that, apart from creating enabling conditions for innovation through investments, structures, and culture, they also need to build the actual innovation capabilities of their companies and employees. To create a mindset for innovation, some BGs have undertaken 'learning missions and trips' to leading international companies at the forefront of innovation, like Apple, Google, Microsoft, Intel, 3M, GE, and Nissan. To further ingrain this mindset, some BGs have begun implementing rigorous internal training and workshops on innovation, conducted by leading experts from around the world. Apart from in-class learning, these programmes and interventions involve a lot of learning through real-time fieldwork, in-company projects, and the like. The idea is to apply what is learnt conceptually as a way to ensure that people develop the practical skills and processes to make innovation happen more effectively and frequently. As part of this, one large BG has also created an international 'boot camp' for senior executives, wherein they have to work in small, frugal teams and go to distant places to develop innovative ideas and business models, then enter and compete in these new markets.

Several BGs have also gone a step further, to actually 'codify' some of the lessons and best practices on innovation, creating a set of usable processes, templates, and tools to manage different stages of an innovation initiative from ideation to experimentation, planning, and execution. While the tools serve as a useful checklist and aid, the process and journey of making these tools provides a big benefit in itself, as the people involved in that effort are able to better reflect upon the key learnings and imbibe them internally. In order to calibrate its baseline for innovation before it starts this capability-building process, the Tata Group has also developed a tool called the 'Innometer' based on the work of some academics. Its goal is to understand the 'current status' of any company's innovation culture and capabilities, identify potential gaps, and thus determine the specific actions and learning interventions required to build the company's innovation capability to the desired level.

Generating External Innovation

Indian BGs have a vast pool of potential talent and resources within their companies to generate innovation over time. Yet, many also realize that a lot of innovative activities lie external to their firms and therefore, they are taking numerous steps and approaches to access such innovation. Clearly, traditional inorganic modes like acquisitions and alliances are critical in this respect. Indian BGs have been quite aggressive in carrying out overseas acquisitions since the turn of the century. From less than USD 5 billion worth of overseas acquisitions in 2003, the activity reached over USD 20 billion by 2013, peaking in 2007 at USD 35 billion just before the global financial crisis (Grant Thornton India 2013). While some of these acquisitions have been done with the objective of seeking entry into those markets for future growth, many Indian BGs have also acquired companies in developed economies to access their technology and talent, in order to generate new products and ideas and upgrade their own internal capabilities here in India. Mahindra's acquisition of SsangYong Motors in South Korea, Tata Motors' acquisition of Jaguar Land Rover and Daewoo, RPG Group's acquisition of SAE Towers, and Sun Pharma's acquisition of Taro Pharmaceuticals are examples.

Indian BGs have also been cautious in how they integrate such acquisitions, taking a somewhat 'hands-off approach' to managing them (instead of giving them a tight bear hug) so that they retain the innovative talent and resources so necessary for future innovation (Kale, Singh, and Anand 2009). Apart from acquiring overseas companies, Indian BGs have also created a slew of alliances and joint ventures with foreign companies as a means to get access to technology, ideas, and best practices for innovation. However, as many of these early joint ventures did not work out as expected, Indian BGs have become a bit more cautious (Kale and Anand 2004).

In recent years, Indian BGs have also been forming new kinds of partnerships for innovation. Some have begun formalizing partnerships with universities for access to cutting-edge resources; for example, Tata Industries' partnership with Tel Aviv University or TCS's partnerships with institutions like Stanford and Massachusetts Institute of Technology (MIT) as part of its Co-Innovation (COIN) programme. Taking inspiration from the concept of open innovation (Chesbrough 2003), and Proctor & Gamble's (P&G) Connect and Develop innovation initiative (Houston and Sakkab 2006), some Indian BGs have also set up networks for crowd-sourcing innovative ideas from the broader community. While both these forms of partnering are widely seen amongst the US companies, they are relatively new for Indian companies.

Innovation in Indian BGs: Challenges and Next Steps

Actions along each of the individual dimensions discussed earlier are critical in driving innovation. The BGs may need to pull levers across all of them so that, collectively, they align to produce innovation.

While the conceptual rationale underlying the potential use of these mechanisms is understandable, the question is whether they will produce the desired results in the long run. Several concerns and challenges lie ahead of Indian BGs as they undertake this journey. The first is the hierarchical nature of decision-making in many Indian organizations. In general, concentrating responsibility for larger strategic moves at the very top of the organization means that leaders cannot take full advantage of the ideas bubbling amongst the younger ranks of decision makers. Second, there is a tendency to give lip service to innovation as a way to respond to stock analysts and external observers. All firms need to show some evidence of a drive to innovation; and there is a danger that traditional members of BGs might discuss innovation-related initiatives without putting real resources and managerial commitment behind them.

Nevertheless, as Indian organizations experience more competition from global firms and as they seek access to global markers, they focus more on innovation as a way to remain relevant to consumers. Indian consumers are highly discriminating in determining the value of products and services and are more than willing to adopt products of firms from other geographic markets, as we see in the automobile industry. Additionally, the BGs that have flourished are those have innovated and remained relevant to local and global markets. The innovation priorities and practices described in this chapter are likely to further contribute to performance differences between firms and business groups.

References

Berger, P. and E. Ofek. 1995. 'Diversification's Effect on Firm Value', *Journal of Financial Economics*, 37: 39–65.

Bertrand, M., P. Mehta, and S. Mullainathan. 2002. 'Ferreting out Tunnelling: An Application to Indian Business Groups', *Quarterly Journal of Economics*, 117(1): 121–48.

Carney, M., E.R. Gedajlovic, P.P.M.A.R. Heugens, M. Essen, and J.H. Oosterhout. 2011. 'Business Group Affiliation, Performance, Context, and Strategy: A Meta-analysis', *Academy of Management Journal*, 54(3): 437–60.

Chacar, A. and B. Vissa. 2005. 'Are Emerging Economies Less Efficient? Performance Persistence and the Impact of Business Group Affiliation', *Strategic Management Journal*, 26(10): 933–46.

Chang, S.J. and J. Hong. 2000. 'Economic Performance of Group-affiliated Companies in Korea: Intragroup Resource Sharing and Internal Business Transactions', *Academy of Management Journal*, 43(3): 429–48.

Chesbrough, H. 2003. *The Era of Open Innovation*. Cambridge, MA: Harvard Business Press.

Chittoor, R., P. Kale, and P. Puranam. 2015. 'Business Groups in Developing Capital Markets: Towards a Complementarity Perspective', *Strategic Management Journal*, 36(9): 1277–96.

Grant, R.M. 1996. 'Toward a Knowledge-based Theory of the Firm', *Strategic Management Journal*, 17 (winter special issue): 109–22.

Guillen, M.F. 2000. 'Business Groups in Emerging Economies: A Resource-based View', *Academy of Management Journal*, 4(3): 362–80.

Houston, L. and M. Sakkab. 2006. 'Connect and Develop', *Harvard Business Review*, 84(3): 58–66.

Kale, P. and J. Anand. 2004. 'The Decline of Emerging Economy Joint Ventures: The Case of India', *California Management Review*, 48(3): 62–76.

Kale, P., H. Singh, and A. Raman. 2009. 'Don't Integrate your Acquisitions, Partner with Them', *Harvard Business Review*, 87(12): 109–15.

Keister, L.A. 1998. 'Engineering Growth: Business Group Structure and Firm Performance in China's Transition Economy', *American Journal of Sociology*, 104: 404–40.

Khanna, T. and K. Palepu. 2000. 'Is Group Affiliation Profitable in Emerging Markets? An Analysis of Diversified Indian Business Groups', *Journal of Finance*, 55(2): 867–91.

Khanna, T. and J.W. Rivkin. 2001. 'Estimating the Performance Effects of Business Groups in Emerging Markets', *Strategic Management Journal*, 22(1): 45–74.

Khanna, T. and Y. Yafeh. 2007. 'Business Groups in Emerging Markets: Paragons or Parasites?' *Journal of Economic Literature*, 45(2): 331–72.

Kogut, B. and U. Zander. 1992. 'Knowledge of the Firm, Combination Capabilities and Replication of Technology', *Organization Studies*, 3: 383–97.

Mahmood, I.P. and W. Mitchell. 2004. 'Two Faces: Effects of Business Groups on Innovation in Emerging Economies', *Management Science*, 50(10): 1348–65.

Morck, R. 2009. 'The Riddle of the Great Pyramids', *NBER Working Paper Series*, WP No 14858.

O'Reilly, C.A., and M.L. Tushman. 2004. 'The Ambidextrous Organization', *Harvard Business Review*, 82 (4): 74–81.

Radjou, N., J. Prabhu, and S. Ahuja. 2012. *Jugaad Innovation*. San Francisco: Jossey-Bass.

Williamson, O.E. 1985. *The Economic Institutions of Capitalism: Firms, Markets, Relational Contracting*. New York: MacMillan Free Press.

Yiu, D.W., Y. Lu, G.D. Bruton, and R.E. Hoskisson. 2007. 'Business Groups: An Integrated Model to Focus Future Research', *Journal of Management Studies*, 44(8): 1551–79.

5

CHIRANTAN CHATTERJEE
SHREEKANTH MAHENDIRAN

From 'Pharmacy' to 'Laboratory'

The Global Biologics Revolution and the Indian Biopharmaceutical Industry

This chapter investigates whether Indian biopharmaceutical firms are equipped to respond to shifts in the global industry as technological change impacts the way drugs are produced. For much of the early twentieth century, the biopharmaceutical industry was embedded in a traditional trial-and-error method of drug discovery that matched disease targets in the body to chemical compounds. This meant that the capabilities of firms in the industry were predominantly oriented towards chemistry-based pharmacology rather than using fundamental biology to understand diseases and uncover cures.

The slow change for the industry started in the 1950s, when science made significant progress in understanding the 'molecularization of medicine' (Pauling et al. 1949). A seminal discovery in this regard was the double helical structure of deoxyribonucleic acid (DNA)[1] and

---

[1] The DNA is a molecule that encodes the genetic instructions used in the development and functioning of all known living organisms and many viruses.

a hypothesis about the mechanism of gene duplication (Watson and Crick 1953). Another key discovery was cell-based mass-production methods of recombinant DNA (rDNA)[2] and monoclonal antibodies—biotechnologies that promised accompanying efficiencies in drug production (Cohen et al. 1973). These advances, referred to in the industry as the 'biotechnology revolution', paved the way for a rational drug design approach and a fundamental shift towards biology-based large molecule drugs.[3] Drug development efforts have switched from a laborious random compound-screening method to one in which molecular biology plays a key role in the discovery and commercialization of new products. Several blockbuster drugs today include this new generation, and innovator firms are switching to biologics-based large molecule

[2] The idea of rDNA was first proposed by Peter Lobban, a graduate student in the biochemistry department at Stanford University Medical School. The first publications describing the successful production and intracellular replication of rDNA appeared in 1972 and 1973. Stanford University applied for a US patent on rDNA in 1974, which was awarded in 1980. The rDNA molecules are DNA molecules formed by laboratory methods of genetic recombination (such as molecular cloning) to bring together genetic material from multiple sources, creating sequences that would not otherwise be found in biological organisms. They are thus also known as 'chimeric DNA', because they are usually made of material from two different species, like the mythical chimera.

[3] The term 'large molecules', for our purposes here, refers to drugs built out of large protein sequences, like biotechnology-based drugs, also known as biologics. Industry watchers classify these into first- and second-generation biologics. Large molecule reformulations or extensions with better efficacy are known as 'biobetters'; and the generic versions are 'similar biologics' or 'biosimilars'. The word 'generic' is avoided, since establishing similarity in large molecules, in contrast to chemistry-based small molecules, requires more than just proving bioequivalence, which is the United States (US) Food and Drug Administration (FDA) criterion for selling generic versions of an innovator small molecule drug. As the 'Supply-side Analysis' section of this chapter will document, large molecule biosimilars, in contrast to small molecule generics, almost always require country-specific clinical trials alongside an array of complex biomanufacturing requirements that make producing them more difficult than producing generic small molecules.

drugs from chemistry-based small molecule products (Branstetter, Chatterjee, and Higgins 2014; Kapoor and Klueter 2013).[4]

This shift in the nature of innovation in the global biopharmaceutical industry has profound implications for catch-up firms that reverse engineer branded drugs developed by innovator firms. These firms were historically adept in producing imitations under the previous regime of chemistry-based new molecules.[5] Notable amongst such catch-up firms was a cohort belonging to the Indian pharmaceutical industry that has played a key role in preserving the balance between escalating drug costs and public access to healthcare. Cipla Ltd, a Mumbai-based firm, disrupted industry dynamics by producing a generic version of an innovator drug for acquired immunodeficiency syndrome (AIDS) at a fraction of the cost (Campaign for Access to Essential Medicines 2011). Cipla is, however, not the only Indian success story in reverse-engineering drugs; and India is now labelled the 'pharmacy of the world' (Bajaj and Pollack 2012). The question is whether this 'pharmacy of the world' will be able to adapt to the changing nature of innovation in the global biopharmaceutical industry and produce large molecule drugs or biosimilars.

This question has important public policy implications because, starting from 2013, many biology-based large molecule products and innovator drugs that make use of them are going off-patent. Industry reports have projected that in 2015, biological drugs will represent some USD 10 billion in annual sales and constitute one-third of drugs coming off-patent[6]—a significant market worthy of attention. But

---

[4] Other examples include drugs to treat rheumatoid arthritis and cancer and generic versions of blockbuster brands, like Gardasil from Merck; Enbrel and Aranesp from Amgen and its partners; Remicade from Johnson & Johnson and its partners; Rituxan, Avastin, and Herceptin from Roche; Lovenox and Lantus from Sanofi-Aventis; and Humira from Abbott Laboratories.

[5] Interestingly, the economics and management literature has documented this evolution of technological change across all industries, not just in biopharmaceuticals, and has discussed its implications for managers and policymakers (Arora and Gambardella 1994).

[6] This is, in fact, a conservative estimate. In China alone, analysts indicate that biologics market would be of this size by 2015, global market size is only

producing biosimilars requires more than what traditional chemistry-based drugs have required to date. Some of these requirements pertain to institutional changes and others to product- and firm-specific capabilities. For one, there is the issue of the regulatory pathways through which biosimilars can be launched in the advanced economies; in the US, for example, this is still unresolved, though the market is gearing towards an entry pathway similar to that of the European Union (EU). Two, producing biosimilars requires more knowledge in basic science and biomanufacturing. The processes are fragile and sensitive and establishing efficacy, similarity, and stability of the biological compounds requires more fundamental work than is required with traditional small molecule drugs. Three, physicians, who are the key change agents and influencers on the demand side, have been reluctant to prescribe biology-based large molecule drugs (Alliance for Safe Biologic Medicines 2012). Finally, producing and selling biosimilar compounds in advanced economies would require non-trivial investments in complementary assets (Teece 1986)—sales and marketing, for example—that might present an entry barrier for generic firms from the Indian biopharmaceutical industry.

All this raises the possibility that firms geared towards small molecule chemistry-based production will need to make concomitant investments and operationalize radical changes to their ways of conducting business. Will they be able to do so? This chapter investigates this question, using a mixed-methods approach that includes secondary analysis of patents and publications of Indian biopharmaceutical firms in the large molecule space. The analysis supplements this with firm-level interviews to uncover the challenges facing the industry in the area of large molecule production capabilities. Finally, with some novel data on the demand side in local markets in India, the analysis provides some initial evidence of heterogeneity amongst firms endowed with variations in capabilities for large molecule production in Indian local markets that could be reflected in their long-term success in exporting.

---

expected to be higher. See https://www.bio.org/sites/default/files/files/BIO_RDAC_Biologics_White_Paper_Jan_2013.pdf.

These findings should be viewed in a nuanced and holistic fashion. At one end, this chapter will seem like a study of failure, with the analysis portraying a lack of innovative and scientific capabilities amongst Indian biopharmaceutical firms in switching to large molecule production from traditional, chemistry-based small molecule compounds. At the other end, however, the analysis of local markets shows that no less than 50 Indian biopharmaceutical firms have already made investments in the production of large molecule drugs, which augurs well for the industry from an export perspective. Nowhere else in the world will one witness such an agglomeration of firm activity in the large molecule biosimilar space. While this is a potential indicator for future industry consolidation and entrepreneurial new-firm formation, in various parts of the value chain and across different therapeutic product markets, it could imply a long-term capabilities explosion for the industry, coming through a division of inventive labour and translating not just to imitation but to innovation in large molecule drugs.

The next section assumes that the reader recognizes the previous existing literature and studies that have explored Indian biopharmaceutical capabilities in generics production.[7] The supply-side analysis here explores the innovative and scientific capabilities of key players in the industry through an assessment of their patenting and publication data. This is complemented with firm-level interviews about the capabilities of Indian biopharmaceutical firms in the large molecule space. The subsequent section provides a broader analysis of these firms' ability to sell large molecule drugs in the local Indian market, with novel aggregate regional demand data for 20 large molecule drugs between 2008 and 2012. This is backed up with an appendix at the end of the chapter that provides an econometric analysis of firm capabilities in local markets of India in large molecule products. The last section concludes with policy and managerial recommendations based on the overall analysis.

Supply-side Analysis

This section consists of a primary and secondary analysis of capabilities in the large molecule space for Indian biopharmaceutical firms,

---

[7] For a detailed analysis, see Chatterjee (2011) and Chaudhuri (2005).

investigating patenting and publication trends before turning to exploratory, qualitative interviews with eight Indian biopharmaceutical firms.

Patenting and Publication Analysis

Using a list of biosimilars launched by various firms in India up to late 2012, we used the firm names to retrieve patenting information from the US Patent and Trademark Office (USPTO), European Patent Office (EPO), and World Intellectual Property Organization (WIPO).[8] We then conjectured that these patents were potentially large molecule related, so long as any of the listed products or related terms featured in the title or abstract of the patent or the publication. The matching process was cross-validated with faculty in biotechnology at the Indian Institute of Technology (IIT) Kharagpur.[9] We present our data in Figures 5.1A, 5.1B, 5.1C, and 5.1D.

We arrived at Figure 5.1A by cross-checking information from company annual reports, discussions with industry experts, and Internet sources indicating that Biocon and Reliance Life Sciences, based in Bengaluru and Mumbai respectively, are the key players in terms of their large molecule programmes within the industry.[10] Each had at least five

[8] Our focal firm cohort for this analysis included Biocon, Reliance Life Sciences, Shantha Biotech, Wockhardt Ltd, Claris Lifesciences, Emcure Pharma, Intas Pharmaceuticals, Cipla Ltd, Dr. Reddy's Laboratories, and Daiichi–Ranbaxy. In addition, because the approval data for similar biologics on the website of India's Central Drugs Standard Control Organization was incomplete and dynamically changing, we had to switch to industry reports. Our baseline for the sample firms and molecules comes from *Generics and Biosimilars Initiative Journal* (2013).

[9] Admittedly, such a matching technique assumes away economies of scope—that is, knowledge unrelated to large molecule patents could be relevant for large molecule innovation—but we leave this point to be addressed by future work. We are grateful here to a student of IIT Kharagpur, Anchit Singh Sekhon, for his research assistance on this matter.

[10] The statistics for Figure 5.1A are based on *Generics and Biosimilars Initiative Journal* (2013), but they are constantly evolving. An interviewee reminded us of this, noting that only a year after our interview, Intas Pharmaceuticals already

to seven ongoing programmes as per the information available at the end of 2012, closely followed by four programmes each at Wockhardt, Intas Pharmaceuticals, and Dr. Reddy's Laboratories. Shantha Biotech, Claris Lifesciences, Ranbaxy, and Emcure Pharma have less than four programmes each. There are around 50 Indian firms operating in the large molecule space, mostly making biosimilars. Figure 5.1c shows the yearly US patent approvals for our focal firms in the large molecule space. Only about 31 patents in total could be matched to large molecule programmes, starting around 2000 and peaking in 2005. In no single year during our period of analysis did the number of US patent approvals exceed 10 in total across all the firms.

Figure 5.1B provides the average duration of US patent approvals in months, measured from day of application to day of approval. This number ranged from six months to just over a year for all our sample patents, but was highest for the US patents approved in 2001, 2008, and 2010. The duration decreased in the last three years of our sample period to around six months, perhaps suggesting that patents are less complex and more incremental in nature as applied by our focal cohort of firms. This conjecture has to be tempered, however, and regulatory learning controlled for, as well as learning by the firms on how to deal with large molecule patents and their concomitant intellectual property (IP) capabilities in appropriating rents from those research and development (R&D) efforts.

So as not to be biased by USPTO data, we next turn to patents retrievable from the EPO and WIPO (Figure 5.1D). Biocon continues to lead the cohort, as Figure 5.1D indicates, with Wockhardt, Dr. Reddy's Laboratories, Intas Pharmaceuticals, Reliance Life Sciences, and Ranbaxy following closely in terms of absolute numbers. Over the years, Indian firms seem to have filed only about 29 WIPO patents and 33 EPO large molecule patents.

---

had eight approved products in the market and more in development, though here we cite the number as fewer than four. He also reminded us that Biocon has four biosimilars and two novel antibodies approved and in the market and five under development.

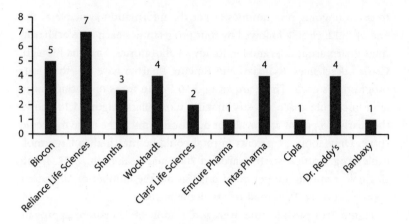

FIGURE 5.1A   Biosimilars by Indian Firms

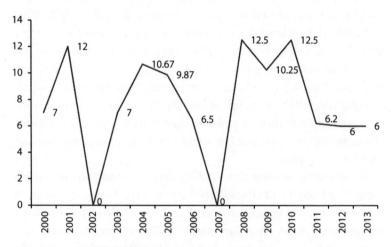

FIGURE 5.1B   Average Duration of US Patent Approval (in months)

Moving over to an analysis of inventor configuration in the US patents (Figure 5.1E), we found that on an average in a year, the number of inventors who filed patents that were approved was about 3.35, with the maximum number of average inventors being seven for approved patents in 2006 and 2009. The number of average inventors per patent approved went up in the middle of our sample period but declined thereafter. This could suggest an increase in spillovers and scientific

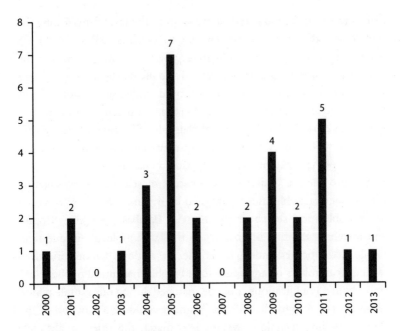

FIGURE 5.1C   Year-wise US Patent Approval of Biosimilars

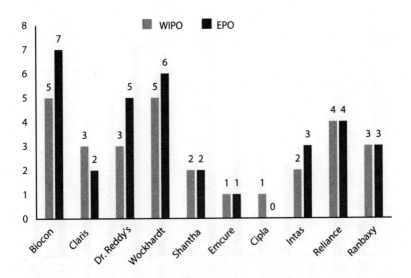

FIGURE 5.1D   Global Patents by Indian Large Molecule Players
Source (Figures 5.1A to 5.1D): Authors' own calculation.

collaboration at the industrial laboratories of the focal firms during the middle period of our analysis. Turning to a claims analysis of the US patents approved for our focal firms (Figure 5.1G), total claims over the sample period averaged 32 per year. If we divide this by our average yearly number of the US patents approved, which was about 2.2, this translates to an average of about 14.5 claims per patent. It is worth noting that in 2004, the total number of claims for all patents approved rose to 97, but this number came down substantially in later years. We can infer that perhaps patents are now broader in nature.

We also tried to understand the nature of science by investigating the number of publications published by our focal firms by conducting a publication search at the Elsevier database, SciVerse Scopus, and matching the title or abstract to the large molecule in question. While R&D-focused Indian firms like Dr. Reddy's Laboratories had far higher numbers of aggregate total biopharmaceutical publications, only about 13 publications could be matched during our sample period (Figure 5.1F). Admittedly, this low number could be a function of the direct-matching technique we used to match the titles or abstracts

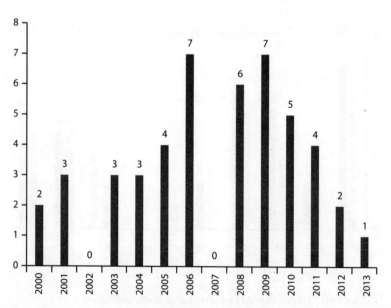

FIGURE 5.1E   Annual Average Number of Inventors in US Patents

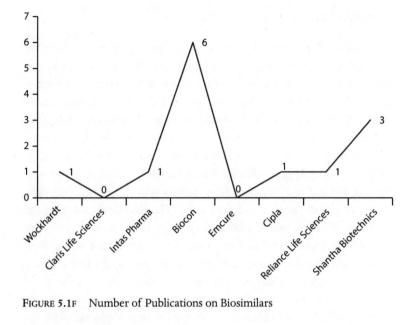

FIGURE 5.1F   Number of Publications on Biosimilars

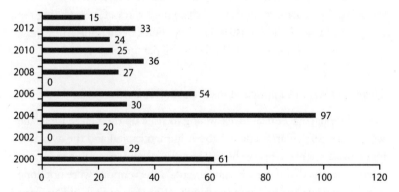

FIGURE 5.1G   Annual Average Number of Claims in US Patents
Source (Figures 5.1E to 5.1G): Authors' own calculation.

of publications to the active ingredient in the large molecule sample. There could be unrelated publications generating scientific knowledge that would be relevant too. Basic large molecule R&D research does, however, seem scanty.

Firm-level Interviews

We conducted qualitative interviews with representatives of eight pharmaceutical firms operating in the large molecule space across India, including founders and top management executives at three firms in Mumbai, three in Bengaluru, one in Pune, and one in Hyderabad.[11] The discussions were with individuals with a scientific, managerial, or techno-managerial background, four of whom held PhDs in areas relevant to biopharmaceutical science. One was a returnee scientist responsible for setting up India's largest biologics and biosimilars programme. The interviews were open-ended and included discussions assessing the large molecule capabilities of the industry and of peer firms. The interviews also explored synergies, if any, between small molecule generics production and large molecule biologics or biosimilars production to get a sense of the firm-level switching costs. The discussions also touched on the firms' overall growth strategy, including views on organic versus inorganic growth and international competitors. We analysed specific aspects of profitability and the capabilities of Indian firms engaged in large molecule work; the regulatory situation in Indian local markets and its importance relative to the global biopharmaceutical value chain; and ways of augmenting the competitiveness of the industry. The key themes emerging from these discussions are summarized in the following paragraphs.

Theme 1: Human Capital and the Supply of Scientists

Indian biopharmaceutical firms aiming to scale up their large molecule work were particularly worried about human capital and the availability of basic scientific researchers. The general understanding from our interviews was that if certain fundamental interventions were not done at a national level to augment the supply of human capital, Indian firms could lose out in the long run to competitors, particularly from Korea, Israel, Malaysia, and even Brazil and Argentina.

One prominent scientist noted, 'We need not just scientists who can engage in basic research but ... who also [have] an eye on the ball

---

[11] Interviewees were promised full confidentiality in the interest of eliciting richer information from the interviews.

in terms of the commercial applications of the R&D programme.' A prominent Bengaluru-based firm's founder added that the clinical development needs of large molecule drugs are distinctly differentiated and needed to be created as a separate segment.

It is worth mentioning the variation in recruitment approaches across firms in this regard. While certain firms have recruited eminent scientists from public research laboratories within India, other firms exhibit different approaches.[12] In general, the ease of mobility of researchers and experts has created positive scientific and research externalities in the industry, especially for early followers in the large molecule space. We also noted a trend of returnee scientists with Western PhDs joining Indian firms' large molecule efforts.[13] A key refrain from scientists was that what they wanted to do with basic industrial research was not aligned with the short-term profit goals of the firms' proprietors. Discussions also pointed to the need for having scientists with an eye on commercially viable 'basic research', rather than basic academic research with uncertain short- and long-term commercial viability. Such a thought aligns with concerns noted in the literature on Pasteur's quadrant (Stokes 1997).[14]

The role of star scientists in enhancing the innovative capacity of biopharmaceutical firms is now well established in the economics of management literature (Azoulay, Graff Zivin, and Sampat 2012; Oettl 2012; Roach and Sauermann 2010). Scientists also aid in establishing what management researchers have called 'component' and 'architectural' competence—thereby enabling the firms to establish product market competencies (Henderson and Cockburn 1994).[15] In this respect, during

[12] For example, M.K. Sahib, currently Director of Genomics and Biotechnology at Wockhardt Ltd—an early mover in the large molecule space—formerly worked for the Central Drug Research Laboratories in Lucknow.

[13] Many subsequently, for example from firms like Wockhardt and Intas Pharmaceuticals, have gone on to other peer firms.

[14] See Kastelle (2013) for an illustration of Pasteur's quadrant as proposed by Stokes (1997).

[15] An example of 'component competence' would be Merck's competence in medicinal chemistry thanks to Max Tishler's leadership, Lilly's competence in diabetic therapy, or Hoffman La Roche's competence in anti-anxiety drugs.

the interviews, the scientists and researchers argued that enhanced innovation productivity is more probable at firms which promote and maintain an extensive information exchange across scientific disciplines and levels within their structures.

The shortage of quality scientists has pushed the industry to start experimenting with innovative self-driven approaches such as 'drug discovery schools' run within firms, as done by Bengaluru-based Advinus Therapeutics. Other options include accessing public research facilities like high-end analytical laboratories to fully characterize different post-translational modifications at the Centre for Cellular and Molecular Platforms (housed within the National Centre for Biological Sciences at Bengaluru). Unfortunately, these are sporadic one-off efforts either by a single firm or within a single location. Clearly, the industry at large needs more investments and policy attention in this regard to aid in the basic science capabilities of Indian large molecule producers.

Theme 2: Competition, Capabilities, and Business Models

The global large molecule pharmaceutical industry is increasingly characterized by stiff competition from various strategic groups, even in the biosimilars segment of the industry. Formally, the industry can be categorized into firms which produce innovative products, biologicals, biobetters, and biosimilars. Even within biosimilars, firms are experimenting with options, opening up competition in the value chain. At one end of the spectrum lie innovator biopharmaceutical firms like Roche, Amgen, and Novartis, which have committed to serious investments in biosimilar programmes themselves and were some of the earliest movers in integrating molecularization approaches into their drug development processes. Firms like Teva and Hospira, established players in the global generics small molecule industry, are

---

An example of 'architectural competence' would be GlaxoSmithKline's experimentation with the Center for Excellence in Drug Discovery Model, which tries to enhance the combinatorial capacities of firms by rejigging the organizational structure and reaping economies of scale and scope from firm-specific component competencies.

also switching to biosimilars. Our interviews suggested that for Indian large molecule players, there might be other threats as well, particularly firms like Celltrion, LG Life Sciences, and Samsung in Korea, as well as a few firms in Malaysia, Singapore, Argentina, and Brazil. That said, the interviews also suggested that no other nation's industry has close to 50 domestic firms operating in the large molecule space. Chinese firms are also a threat, though interviewees agreed that they might be a generation behind in their efforts to scale up their large molecule activity.[16]

Other than potential competitors, Indian biopharmaceutical firms, especially those that did not enter the large molecule space early, might also face a tricky decision on *making* versus *buying* clones. Molecular cloning is the laboratory process used to create rDNA. Formation of rDNA requires a cloning vector, a DNA molecule that will replicate within a living cell. It is in these rDNA principles that the capabilities of innovator or imitator biopharmaceutical firms differ.

Large firms like Amgen and Lilly (which licenced the first drug produced using rDNA, namely, human insulin developed by Genentech) have in-house cloning facilities. Clones are, however, also available from specialized cloning firms, a slew of which exist in Europe and the US, including Selexius, Bioexpress, Rhyme Biotech, Catalent, Lafagen, and Aragen Bioscience. Cloners are in themselves an industry sub-market, offering division of labour in the classic Adam Smith sense. They offer clones for each large molecule programme at something around USD 10,000.[17] Early entrants in the Indian large molecule space like Wockhardt and Biocon have traditionally bought clones to accelerate their entry into the industry. However, newer players, like Mumbai-based USV Ltd, have decided to build in-house cloning facilities, perhaps

---

[16] This might or might not be true, given that Cipla has actually prospected an inorganic route in expanding its large molecule capabilities by buying a 25 per cent stake in BioMab, a Shanghai-based maker of large molecule drugs. It must also be noted, as an interviewee pointed out, that 'a recent survey of 352 global biomanufacturers ranked China as the ultimate destination for outsourced biomanufacturing, with 17 per cent of the respondents identifying it as their top destination. India was the choice for 13.2 per cent of respondents.'

[17] This number might be conservative; some of our interviewees indicated that the costs might be even five to ten times higher.

with an eye on the long run.[18] This is a tricky trade-off but, as a scientist pointed out in the interviews, might be a wise decision.

Whether Indian firms make or buy clones, the issue resonates classically with the idea of markets for technology (Arora, Fosfuri, and Gambardella 2001), which has implications for corporate strategy. It is also worth pondering why Indian players could not also offer specialized cloning services to the world. Interviewees conjectured that this might be because, to be a successful specialized cloner, one needs proprietary technology (like Catalents's high expression technology), capabilities for which might not exist in Indian firms.[19] It might also be a path dependency issue, as postulated by Patel and Pavitt (1997). Relatedly, another interviewee conjectured that Europe has a traditionally strong base in contract manufacturing and that industry may have spun off these cloning firms.

The trade-off in making or buying clones also highlights the likelihood of the increasing propensity for alliances to form in the large molecule space, not only in research but also in downstream development and complementary assets. Already some notable deals have been forged, including those between Ranbaxy and Pfenex (*The Hindu* 2010), Merck Serono and Dr. Reddy's Laboratories (in 2012; India Infoline n.d.), and Biocon and Pfizer.[20] This is not just an Indian phenomenon.[21]

---

[18] Scientists argued that this is because cloners tend to provide clones with an eye on the yield, whereas high-quality clones are required for product stability to establish similarity and meet regulatory standards during clinical trials phase. In-house cloning might be necessary to cater to this discussion on stability and ensuing quality of biosimilars.

[19] There are exceptions here and there. For example, while not exactly developing cloning-related proprietary technology, Bengaluru-based Connexios Life Sciences, at the time of writing this chapter, is trying to build a general purpose technology (Bresnahan and Trajtenberg 1995) to solve diseases better using translational–network biology approaches, and is testing its proprietary technology in the therapeutic markets of type II diabetes and related metabolic diseases.

[20] While this deal was forged in 2010 (see Pfizer 2010), it fell through later in 2012 (see Kulkarni and Foy 2012).

[21] In 2010, for example, Teva and Ratiopharma entered this space with a USD 3.6 billion alliance, followed by another deal in the same year between Johnson & Johnson and Crucell worth USD 2.4 billion. Merck also entered a

One interviewee noted that these deals are inevitable, since succeeding in biosimilars 'is not a sprint but a marathon'. He further noted that, to be successful in global markets, firms need to invest in complementary assets like 'specialized sales and distribution representatives (costing in excess of USD 200,000 in annual salary per person), who can go to doctors with not just one product but a portfolio of large molecule products.' It is worth noting here that the structure of the global industry also creates entrepreneurial opportunities for Indian players to branch out into specialized areas of activity, like offering contract biomanufacturing services or niche products like large molecule vaccines.[22]

Theme 3: Regulation, Firm Focus, and the Institution-based View of the Firm

The third key theme is the regulatory threshold in the large molecule space, especially in biosimilars. This is especially important depending on whether the firm focuses on local or overseas markets. Even outside the EU (which has the only existing regulatory entry pathway for large molecule biosimilar products, though the US is expected to mimic it), one issue is increasingly becoming clear (Grabowski et al. 2006): each local market will require local trials at the Phase I and Phase III levels, at least, which highlights the high entry barriers for generic biological products.

In India, for example, clinical trials are a sensitive issue, with the Supreme Court stalling clinical trials that failed to comply with an elaborate three-tiered review mechanism in 2012.[23] One interviewee noted

---

strategic alliance with a Chinese firm, Hanwha, in 2011 for about USD 0.7 billion to scale up its efforts in the large molecule space.

[22] One interviewee indicated, 'Biocon has the largest mammalian cell-culture-based biomanufacturing capacity in India, followed by Reliance Life Sciences, Dr. Reddy's Laboratories, Intas Pharmaceuticals, and Kemwell. In microbial biomanufacturing of recombinant products, Wockhardt and Intas closely follow Biocon, while in vaccine production Shantha Biotech, Serum Institute, Wockhardt, and Panacea Biotech are the largest players.'

[23] *Swasthya Adhikar Manch v. Union of India*, WP (Civil) No. 33/2012, available at http://www.iscr.org/pdf/7-Supreme%20Court%20Order%20-%20Oct%2021,%202013.pdf.

that this issue might 'actually cripple the biopharmaceutical sector, not just for its local market prospects but also for exports, especially when one looks at South Korea, where the regulatory system is expeditious in terms of clinical trials and development even in local markets'. Another interviewee conjectured even further stating that this might mean that 'launching a large molecule in India will now take close to three to four years, from a previous regime of one year, and will involve a doubling or even tripling of costs'.[24] A different regulatory regime related to clinical trials, in fact, benefited early mover firms in large molecules; the followers could suffer from this shift in the institutional environment.[25]

Another key aspect that came out of our discussions pertaining to firms' geographic focus in selling their large molecule products. Given the need for developing expensive complementary capabilities to penetrate advanced economies like that of the US, Indian firms are prospecting emerging markets as viable destination markets for their large molecule products. Our discussions on the capabilities of Indian firms bolstered this. One interviewee indicated that 'clinical development cost per large molecule biosimilar is around USD 100 million; setting up a Good Manufacturing Practices certified biomanufacturing facility will require some USD 150 million; added to which will be the expenditure on clinical trials and on investments in complementary assets for sales, advertising, and promotions'.

[24] Again, we faced some variation in opinion. One interviewee conjectured more plausibly that 'in fact the faster timeline is three to four years, if there are no regulatory hurdles and the technical work goes well; with regulatory hurdles that number could go up substantially. There is no way anywhere in the world to launch a product in one year.'

[25] A classic example here relates to Dr. Reddy's launch of Reditux (ingredient Rituximab) in India in 2007, with smaller sample of patients recruited for clinical trials to treat leukaemia and non-Hodgkin's lymphoma. Another relates to Biocon's launch of a novel biologic drug, Alzumab (ingredient itolizumab) in 2013. Alzumab is claimed to be the world's first anti-CD6 monoclonal antibody for the treatment of psoriasis patients in India, but several industry representatives admitted in conversations with us that Biocon 'were fortunate to avert hold on clinical trials' for this particular drug. For Alzumab, Biocon conducted one of the largest trials in the country ever, on 400 patients, and is planning to launch an investigational new drug application with the US FDA.

An interviewee pointed out that all of this might mean that 'a total investment for a single global programme [could range] from USD 350 million to USD 900 million with a six- to eight-year period of investment'. Indian firms might not have enough deep pockets and patience to focus on such a project, but could actually focus on key emerging economies, notably, China and Brazil. This is an interesting observation since, if it indeed does happen that way, one might see history repeating itself from the past global evolution of small molecule drug sales. In fact, patent battles on large molecules are being fought,[26] à la Glivec, in the coming days, with the Indian regulatory system and firms offering the world a 'patent law 2.0' related to large molecule drugs (Kapczynski 2013).[27] In fact, India has already revoked Roche's patent for its large molecule Hepatitis C drug, Pegasys (Babu 2012) and biosimilars in India might soon enter emerging discussions on reasonable pricing, while being protected with patents in the country (Chatterjee, Cubo, and Pingali 2015).

Beyond geographic focus, firm managers also contemplated the local regulatory framework in launching large molecule products within India and across emerging economies. Several interviewees felt the same, with one of the interviewees remarking, 'India has the potential to lead the way in establishing a novel and abbreviated biosimilars

[26] For instance, there is the litigation between Roche and Biocon/Mylan, before the Delhi high court, over the biologic drug Herceptin, which begun in 2014. See Shanker (2016).

[27] Glivec is a Novartis-patented drug whose patent was not granted in India in 2013 after a prolonged court battle. The key argument was that the innovation behind the drug was not fundamental enough. This ruling spurred a flurry of transnational activity, with media reports conjecturing that other countries might borrow from the Indian patent court's decision and implement similar measures to reject innovators' patent-protected drugs. See Sidhartha (2013). Firm interviewees also shared with us that there are quite a few large molecule products which innovator firms 'forgot to patent' or for which they 'ignored' patenting in Indian markets, such that Indian producers could reverse engineer similar biologic drugs and launch them in local markets. Recent times have, however, seen innovator firms being more careful about their Indian patent strategies. In fact, Herceptin is a biologic by Roche that is undergoing patent battles in India. See Shanker (2016).

pathway based on molecular characterization and high-end pharmaco-kinetic or pharmacodynamics studies.' The current regulatory framework for launch of biologics and biosimilars in India relates to the 'Regulatory Requirements for Marketing Authorization in India', issued by the government's Department of Biotechnology, which closely follows the EU and World Health Organization (WHO) guidelines. But some interviewees felt that this could be further transformed. One interviewee stated that 'Phase III trials could be restricted to demonstrate response-based efficacy, safety, and immunogenicity, especially in contrast to the current biosimilars guidelines that aim to demonstrate non-inferiority based on long-term efficacy rather than response.' This, some interviewees felt, was both expensive and time consuming. In sum, interviewees felt that there is an opportunity for drug regulatory bodies in India to 'form a regulatory consortium with emerging economies to harmonize this approach'.

Whether or not these recommendations will bear fruit shall indeed require careful contemplation. Irrespective of this, the key takeaway is that regulations and institutions are going to play a key role as to how the future of the industry will evolve. The conversations further bolstered an emerging view in the strategy literature that documents the role of the institution-based view of the firm (Ahuja and Yayavaram 2011) and firms' ability to appropriate 'influence rents' over and above standard rents, as investigated in the economics and management literature. This view argues that what differentiates the more successful firms is not industry structure or resources, but their capabilities to influence the institutions they face. One can expect a similar prediction to be borne out for Indian biopharmaceutical firms focused on the large molecule space.

Demand-side Analysis

A key discussion point with firm-level executives was the criticality of local Indian markets in large molecules to understand the capabilities of the firms operating in this space. Accordingly, the demand data of 22 focal molecules manufactured and sold by 70 firms (50 of which are domestic), from 2008 to 2012, across 20 Indian states, varying across seven therapeutic categories, was sourced and analysed from the All India Organisation of Chemists and Druggists (AIOCD).[28] Econometric models were then used to analyse the heterogeneous firms' ability to produce varying quantities of the complicated biopharmaceutical molecules, as well as their capacity to launch product varieties and adopt competitive pricing strategies.[29]

The heterogeneous nature of a firm was measured using three types of classifications to understand the different dimension of a firm's intrinsic character and its impact on market-related outcomes. The first was whether the firm was a domestic or multinational entity. The second classification was on the basis of the timing of market entry: 'early mover' (first entrants in the large molecule space); 'early follower' (which followed the path paved by the early movers); and 'others' (the remaining firms).[30] Third was whether the headquarters of a domestic firm was based in Andhra Pradesh, Gujarat, or anywhere else in India. The rationale for the third classification was based on agglomeration economies and the anchor tenant hypothesis, following the literature in strategy and economics (Agarwal and Cockburn 2003; Shaver and Flyer 2000).

The econometric models exploring the level of influence of a firm's heterogeneity on its ability to manufacture and sell large molecule drugs indicate that domestic Indian firms, on average, are associated with a higher level of aggregate quantity being sold in the large molecules market in India during the sample period.[31] Specifically, on average, domestic firms manufacture and sell 13,332–4,610 more units of the focal brand/stock-keeping unit (SKU) than multinational firms (see

[28] These are the same molecules that were used in the previous section for the purpose of inventive output analysis and discussion. The seven therapeutic categories are drugs for: (a) diabetes; (b) the genito-urinary system; (c) sex hormones; (d) general systemic anti-infectives; (e) blood and blood-forming organs; (f) anti-neoplastics; and (g) the musculo-skeletal system.

[29] See Appendix 5A for details on the data and econometric models.

[30] This classification process was carried out in consultation with our industry interviewees. A detailed classification heuristic is available from the authors on request.

[31] Equation 1, mentioned in Appendix 5A, was used to estimate the model.

Table 5A.4 in Appendix 5A). Further, the regression estimates indicate that early movers manufacture and sell more brands/SKUs (on average, 2,146 units more) relative to the remaining firms (excluding early followers) operating in the large molecule space (see Table 5A.5 in Appendix 5A). This corroborates our managerial interviewees' statements that large molecules and similar products are all about entering early and being first movers in the game, even for loosely regulated markets like India.

Further, our analyses reveal that the early movers have a great ability to launch more brands/SKUs in comparison to early followers and the remaining firms in the market (see Table 5A.8 in Appendix 5A). This finding reiterates that the early movers not only hold an advantage in terms of quantities of brands/SKUs sold, but are also more likely to appropriate economies of scope by launching product varieties, relative to other firms.

The analysis of regional externalities shows that domestic firms in Andhra Pradesh were producing, on average, 3,064–218 units of brands/ SKUs fewer than firms located in other Indian states (see Table 5A.6 in Appendix 5A). While this evidence reveals the existence of region-level externalities in the Indian market, it is indicative of the strategic focus of these firms as well. Firms in Gujarat and Andhra Pradesh, which are the traditional chemistry-based pharmaceutical clusters in India, are potentially choosing to remain focused on operating in the small molecule business, thereby not switching their product focus or strategic groups to large molecules (McGee and Thomas 1986). As a result, one can extrapolate and hypothesize that newer domestic firms might be more successful in the large molecule space.

With respect to prices, we find that the maximum retail price (MRP) was positively related to quantity sold, indicating an upward-sloping demand curve.[32] While this result may be counter-intuitive at first glance, it actually indicates the intrinsic nature of the demand curve of these products. The biopharmaceutical products are complex and sophisticated in nature, thereby potentially inducing consumer

[32] The sign of the coefficient on prices (MRP) is consistent and positive in all the three models estimated using Equation 1. The estimated coefficient is given in Tables 5A.4, 5A.5, and 5A.6 .

behaviour such as with Giffen goods, wherein people consume more of the same good as prices rise contrary to conventional intuition. This could be because the agent of the consumer, the patient, a typical physician prescribing more of the same product despite rise in prices due to clinical requirements.[33] Despite the possibility of Giffen nature of large molecule goods, our findings show that domestic Indian firms charge INR 160—INR 237 (about USD 2.58–USD 3.82 per brand/ SKU, assuming the conversion rate of INR 62 to 1 USD), on average, which is less than the multinational firms for the average focal brand/ SKU (see Table 5A.7 in Appendix 5A). This result could also indicate commoditization of the large molecules in the sample. It should be noted, however, that prices are much higher on average for the large molecule drugs.[34]

Figure 5.2 provides evidence on total revenues accrued from sales of large molecules measured in nominal Indian rupees. In 2008, Karnataka, Tamil Nadu, and Maharashtra were the top three regional markets in terms of revenue.[35] However, the revenue generated in Karnataka fell by 10 per cent, from INR 758 million (about USD 16 million) in 2008 to INR 677 million (about USD 14.4 million) in 2012, using the average exchange rates for USD to INR for the respective years.[36] In 2012, the highest revenue-generating markets were Tamil Nadu, Maharashtra, and Kerala. During our period of analysis, Chhattisgarh, Jharkhand, Bihar, and Odisha were the states whose firms performed consistently poorly in terms of revenue generated. Perhaps the premiums firms

[33] Another explanation for the positive relationship between price and quantity sold might be the problem of identification, for which a more sophisticated analysis must be carried out to pinpoint the direction and magnitude of causality. Future work can explore econometric approaches, such as an instrumental variable approach and structural demand modelling, to estimate the exact relationship between prices and quantity sold of large molecule products.

[34] Industry experts mentioned during the managerial interviews that conventional small molecule drugs are priced higher on average than large molecule drugs.

[35] The revenue of a particular SKU/brand was calculated by multiplying its quantity sold with its MRP.

[36] We assume an exchange rate of INR 47 to USD 1 for the period.

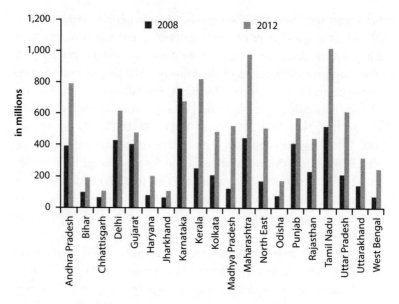

FIGURE 5.2   Revenue Generated during 2008 and 2012 (in INR million)
Source: Authors' own calculation using data sourced from AIOCD for respective years.

could charge in these states were not as high, on average, as those in the richer states in India.

In summary, our analysis provides some very novel and initial evidence on the state of affairs in local Indian large molecule markets and the role of firm heterogeneity therein. We document the crucial role of domestic Indian firms in being able to produce higher quantities than their multinational counterparts. We also document that firms from the domestic cohort which entered the large molecule space early are the ones enjoying the advantage of being first movers, in being able not just to sell higher quantities but also to launch greater product varieties and thereby reaping economies of scope. An interesting finding is the role of new entrants in the domestic cohort coming from firms outside of the traditional pharmaceutical clusters in India, which can sell higher quantities of products in the large molecule space. Domestic firms are also pricing lower on average than multinational firms, which may be constrained to keep prices closer to global levels given that their products could be patented in Western markets.

★★★

Our analysis in the earlier sections offers several novel findings with policy and managerial implications. The global biopharmaceutical industry is undergoing some radical changes induced by innovation in basic-science- and biology-induced discoveries. This has meant that the traditional model of chemistry-based small molecule drugs are giving way to new generations of large molecule biotechnology-driven products. This shift in the nature of innovation in the industry also has profound implications for how catch-up firms will transform, given the shifting contours of the industry. This chapter explores this question with the case of the Indian biopharmaceutical firms, acknowledged to be the 'pharmacy of the world' in small molecule drugs.

Using a mixed-methods approach that combines analysis of patenting and publication data, firm-level managerial interviews, and econometrics-driven empirical analysis of Indian markets for large molecule drugs, the findings in this chapter offer a nuanced portrait. Innovation and basic-science capabilities are weak in the industry. These findings raise a policy imperative to take more concrete steps in this regard, especially in improving the pool of biological sciences talent domestically. Competition from other cohorts of firms outside India, for example, in Korea or Israel, might be particularly worthy of attention. Firms also need to think about investing in complementary capabilities to succeed globally in the large molecule space. Regulatory uncertainty, both locally and abroad, also means that firms' ability to surmount the obstacles posed by institutions will play a critical role in their success. In particular, newer patent battles in the large molecule space may unfold.

The econometric analysis of aggregate regional demand data indicates that domestic firms are able to produce higher quantities and charge lower prices than multinational firms. But even within domestic firms, there is substantial heterogeneity between early movers in the large molecule space and the rest of the cohort (in their ability to produce quantities, price products, or launch product varieties to reap the advantages of economies of scope). Firms also seem to be coming from beyond the traditional clusters of pharmaceutical production, such as Andhra Pradesh and Gujarat, with novel capabilities in the large molecule space.

Overall, our findings can be categorized into three key issues: (a) a firm's ability to circumvent regulatory uncertainty, (b) its ability to upgrade in a techno-scientific fashion to cater to the more sophisticated demands of producing similar large molecule compounds, and (c) its ability to go to market with complementary assets and concomitant investments in capabilities.

At a broader level, these issues recreate history in the global evolution of the biotechnology industry (Moxley 2012). For example, for Pentyde, an anti-allergy drug developed by the San Diego area start-up company Immunetech Pharmaceuticals in the 1970s, regulatory challenges to clearing trials requirements at the US FDA proved a critical barrier, though the drug was based on good science and was ready to go to market with required capability investments. Consider, in contrast, the experience of Leukine, an early rDNA therapeutic drug from one of the biotechnology industry's pioneers, Seattle-based Immunex. The firm, while doing good science and surmounting regulatory barriers, failed to make efficient corporate decisions and did not succeed in the market. Both these examples indicate that a robust and concomitant convergence of the three issues, good science, wise regulation, and rational corporate investment in complementary assets, will crucially decide the future prospects of Indian biopharmaceutical firms in the large molecule space. Being able to surmount this troika of issues could aid in converting the 'sprinters' to 'marathoners' and help transform the small molecule 'pharmacy of the world' into its large molecule 'laboratory' as well.

Appendix 5A: Note on Data and Empirical Framework

The AIOCD is the nodal industry organization for chemists, both retailers and wholesalers, in India, with a membership exceeding 0.75 million. We procured data made available by the AIOCD on large molecule aggregate sales through private chemists and pharmacists in India for our focal molecules identified earlier. The data contained aggregate regional demand of 70 firms across 20 Indian states for 22 large molecules (see Table 5A.1) in our sample, over five years between 2008 and 2012. The raw data also provided information on sales for 592 large molecule brands or SKUs in our sample, after cleaning for missing data.

A general descriptive analysis of the demand data shows that the number of firms that competed in Indian states has increased (see Table 5A.2),

TABLE 5A.1   List of Molecules Used for Demand-side Analysis

Large Molecule Name	Therapeutic Market Description
Fast-acting, soluble, or neutral insulin	Anti-diabetic/alimentary tract and metabolism
Follicle-stimulating hormone	Genito-urinary system and sex hormones
Hepatitis B	General systemic anti-infectives
Human chorionic gonadotropin	Genito-urinary system and sex hormones
Darbepoetin alfa	Blood and blood-forming organs
Interferons, beta	Anti-neoplastic and immunomodulating agents
Plain human insulins	Anti-diabetic/alimentary tract and metabolism
Streptokinase	Blood and blood-forming organs
Epoetin alfa	Blood and blood-forming organs
Erythropoietin products	Blood and blood-forming organs
Etanercept	Musculo-skeletal system
Filgrastim	Anti-neoplastic and immunomodulating agents
Insulin devices	Anti-diabetic/alimentary tract and metabolism
Pegfilgrastim	Anti-neoplastic and immunomodulating agents
Reteplase	Anti-diabetic/alimentary tract and metabolism
Fast-acting, insulin lispro	Anti-diabetic/alimentary tract and metabolism
Interferon Alfa 2a	Anti-neoplastic and immunomodulating agents
Interferon Alfa 2b	Anti-neoplastic and immunomodulating agents
Other human insulins	Anti-diabetic/alimentary tract and metabolism
Pegylated interferon Alfa 2a	Anti-neoplastic and immunomodulating agents
Pegylated interferon Alfa 2b	Anti-neoplastic and immunomodulating agents
Rituximab	Anti-neoplastic and immunomodulating agents

Source: Generics and Biosimilars Initiative Journal (2013).

TABLE 5A.2   Number of Firms by States: 2007–13

States	2007	2008	2009	2010	2011	2012	2013
Andhra Pradesh	21	28	32	37	31	28	15
Bihar	15	25	29	30	23	24	13
Chhattisgarh	21	21	24	24	22	21	11
Delhi	21	26	25	25	32	26	19
Gujarat	30	32	34	39	34	31	16
Haryana	25	27	30	25	29	31	14
Jharkhand	17	18	21	27	27	23	11
Karnataka	26	20	28	30	34	29	17
Kerala	24	33	28	27	37	30	25
Kolkata	17	29	29	27	36	27	19
Madhya Pradesh	14	21	25	28	36	33	19
Maharashtra	28	36	28	33	32	34	23
North East	19	24	26	26	31	29	15
Odisha	16	23	29	32	25	20	14
Punjab	28	34	33	35	31	36	15
Rajasthan	22	35	31	27	37	31	18
Tamil Nadu	24	33	21	30	34	26	18
Uttar Pradesh East	24	33	32	28	38	30	19
Uttarakhand and Uttar Pradesh West	27	27	34	34	34	34	20
West Bengal Rest	18	30	21	25	31	24	17
Average	22	28	28	29	32	28	17

Source: Authors' own calculation using data sourced from AIOCD for respective years.

on average, from 22 every month in 2007 to 28 in 2012. Gujarat, Haryana, Maharashtra, Punjab, Rajasthan, Uttar Pradesh, and Uttarakhand have a higher concentration of firms selling their products in the market. Further, Madhya Pradesh and the north-eastern states have seen a higher growth presence. For example, the number of firms in the large molecule space in Madhya Pradesh has increased to 33 in 2012 from 14 in 2007. While all states have witnessed an increase in firm presence, the growth rate in rich, developed states in India has been nominal in comparison to less- or least-developed Indian states, perhaps indicating greater penetration of large molecule sales in poorer regions of India.

The total quantity of large molecule products sold in the local market has increased substantially from 2007 to 2012 (see Table 5A.3). On average, the quantity sold increased by 207 per cent from 2007 to 2012. In particular, Kerala witnessed the highest growth, with the aggregate number of brands/SKUs sold increasing from 323,355 in 2007 to 3,541,212 by 2012. The surge in quantity sold has been mainly due to the quantity sold by multinational firms sky-rocketing from 189,558 in 2007 to 3,028,102 in 2012. This is in addition to the fact that the quantity sold by the domestic firms doubled during the same time period in Kerala. The north-eastern states had the second-highest growth rate, about 465 per cent, in terms of quantities of brands/SKUs sold. It is, however, to surprising to find that Karnataka witnessed a negative growth rate, about 19 per cent, even though it was the fifth-highest state in terms of absolute quantities sold amongst the 20 Indian states in 2012.

The top three states were Andhra Pradesh, Maharashtra, and Tamil Nadu, which witnessed a doubling of total quantities of brands/SKUs sold during the sample period. Significant increases are found in the bottom three states, where sales tripled by the end of 2012. The surge in quantity in all the 20 sample states indicates a vibrant market for large molecule products in India.

In terms of therapeutic markets, anti-diabetic and genito-urinary system drugs and sex hormones contribute about 72 per cent and 20 per cent respectively to the total quantity of brands/SKUs sold across all states and years (see Figure 5A.1). Only about 47,269 units of drugs related to the musculo-skeletal system SKUs were sold.

The following sections explain the empirical framework adopted to measure the firms' ability to manufacture and sell large molecule drugs and vary their pricing strategy and capacity to launch different products in the Indian regional markets.

Empirical Models for Quantity Sold

Equation 1 follows past work in the literature on biopharmaceutical economics (Danzon and Chao 2000) that models the quantity Q, that is sold by a firm i, in an Indian state-level region s, for a molecule

TABLE 5A.3 Sales of Large Molecule Drugs across India: 2007–12

States	2007	2008	2009	2010	2011	2012	Per Year Growth Rate (2007–12)
Andhra Pradesh	958,621	1,659,232	1,888,338	2,576,942	3,234,114	2,563,808	27.91
Bihar	230,981	269,751	375,776	538,607	640,466	624,887	28.42
Chhattisgarh	148,953	203,665	291,696	321,138	388,073	346,364	22.09
Delhi	796,989	1,118,152	1,100,976	1,209,967	1,916,842	2,003,468	25.23
Gujarat	817,820	1,166,163	1,336,480	1,357,029	1,518,025	1,576,557	15.46
Haryana	179,228	265,748	322,395	367,590	523,127	484,165	28.36
Jharkhand	206,288	212,559	232,326	306,190	435,544	322,052	9.35
Karnataka	2,614,755	3,957,974	3,240,108	3,413,471	1,849,280	2,118,674	–3.16
Kerala	323,355	468,150	492,803	829,289	3,390,473	3,541,212	165.86
Madhya Pradesh	344,732	554,569	623,443	650,359	777,953	886,335	26.18
Maharashtra	1,408,408	2,168,840	2,460,658	2,924,786	3,474,692	3,365,518	23.16
North East	191,885	269,721	339,203	522,841	962,651	1,085,328	77.60
Odisha	203,272	388,296	427,349	477,052	697,934	817,083	50.33
Punjab	491,041	665,140	659,138	690,312	777,790	854,152	12.32
Rajasthan	563,267	784,786	725,678	764,333	1,026,773	1,038,497	14.06
Tamil Nadu	1,216,575	1,917,729	2,893,984	3,448,096	4,039,623	4,257,274	41.66
Uttar Pradesh East	378,379	546,099	547,292	908,541	804,766	771,412	17.31

Uttarakhand and Uttar Pradesh West	336,842	499,077	655,865	758,709	746,889	852,307	25.50
Kolkata	401,540	717,470	818,278	1,256,785	1,105,632	1,205,433	33.37
West Bengal Rest	108,885	146,641	168,152	189,493	360,674	420,438	47.69
Total	11,520,276	17,262,292	18,781,660	22,254,745	27,565,689	27,929,531	23.74
Average	596,091	898,988	979,997	1,175,577	1,433,566	1,456,748	24.06

Source: Authors' own calculation using data sourced from AIOCD for respective years.

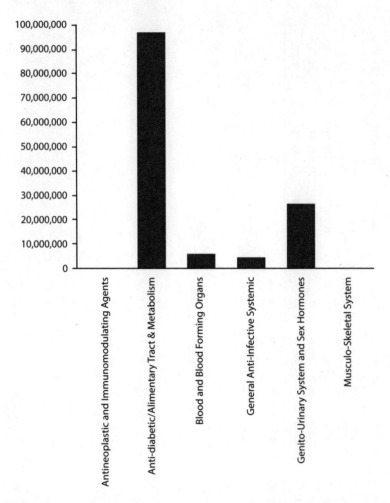

Figure 5A.1 Total Quantity of Sales of Large Molecules across Therapeutic Markets
Source: Authors' own calculation using data sourced from AIOCD.

m, and for a brand/SKU b at the yearly level t.[37] The econometric specification will help us understand the role of firm heterogeneity

---

[37] Note that Q is essentially the total quantity of various brands/SKUs sold by firms and does not adjust for the strength of the medicine in the focal SKU. This could create some biases in our inferences, but since most large molecules

on its capability to manufacture and sell large molecule drugs across 20 Indian states.

$$Q_{i, s, m, b, t} = \alpha_0 + \beta_1 (\text{Dummy for Firm Heterogeneity}_i)$$
$$+ \beta_2 (\log \text{ of price}_{b, t}) + \alpha_s + \mu_m + \beta_b + \tau_t + \alpha_s \times \tau_t$$
$$+ \varepsilon_{i, s, m, b, t} \tag{1}$$

Our variable of interest on the right-hand side of the econometric specification is the 'dummy for firm heterogeneity', which we try to measure in one of the three ways: (a) whether a firm is a domestic or multinational entity; (b) whether the firm is an early mover, late mover (early follower), or other in the large molecule space; and (c) whether, amongst the domestic cohort, the firm is based in Andhra Pradesh, Gujarat, or anywhere else in India.

Equation 1 also controls for brand-level prices with log of prices. Several other factors like molecule-specific idiosyncrasies, along with unobserved heterogeneity due to brand-level preferences, yearly changes, and state-level unobserved heterogeneity that are time-varying or are time-invariant, could confound the causal inference of how firms are heterogeneous in their ability to sell varying quantities of brands. To address these, we use state-level, molecule-level, brand-level, year-level, and state × year-level fixed effects in our econometric strategy, denoted by $\alpha_s, \mu_m, \beta_b, \tau_t, \alpha_s \times \tau_t$ in Equation 1. The sign and significance of β_1 will help us understand how firms are heterogeneously abled to produce varying quantities of products.

These ordinary least squares (OLS) regressions are estimated using an unbalanced panel from the AIOCD data and using Equation 1. The dependent variable is actual quantity of units of brands/SKUs sold. The specifications control for fixed effects at various levels to account for unobserved heterogeneity. Tables 5A.4, 5A.5, and 5A.6 provide the regression estimated using Equation 1, exploring the three types of heterogeneity classification of firms mentioned earlier.

---

are sold in the injectable format, we found it difficult to adjust for the strength of the brand; we leave it for future work to take it up.

TABLE 5A.4   Quantity Models—Firm Heterogeneity Based on Domestic/
Multinational Firms

OLS Regressions	Dependent Variable: Actual Quantity Sold	
	Model 1	Model 2
Dummy (1= Indian,	13,332***	14,610***
0 = Multinational)	[1,149]	[370.3]
Logged MRP	1,228***	1,066***
	[239.9]	[210.6]
Constant	−20,965***	−26,503***
	[1,762]	[3,185]
State Fixed Effects	Yes	No
Developed Status#	No	Yes
Molecule Fixed Effects	Yes	Yes
SKU Fixed Effects	Yes	Yes
Year Fixed Effects	Yes	Yes
State × Year Fixed Effects	Yes	No
Developed Status × Year	No	Yes
Units of Observation	Firm–State– Molecule–SKU–Year	Firm–State– Molecule–SKU–Year
Cluster	Firm level	Firm level
Number of Firms	70	70
Number of Molecules	22	22
Number of States	19	19
Number of Years	5	5
Number of SKUs	592	592
Observations	20,793	20,793
R-squared	0.414	0.408

Source: Authors' own calculation using data sourced from AIOCD.
Notes: 1. Robust standard errors are given in brackets.
2. *** $p < 0.01$, ** $p < 0.05$, * $p < 0.1$.
3. # The states have been grouped according to the developed status provided in the Rajan Committee Composite Development Index. In case of Delhi, the committee report does not categorize it under any group; however, we have placed it in the relatively developed category. Further, all north-eastern states have been placed in the less-developed category.

TABLE 5A.5   Quantity Models—Firm Heterogeneity Based on Entry Timing

OLS Regressions	Dependent Variable: Actual Quantity Sold	
	Model 1	Model 2
Firm Type: Followers/	350.7	44.68
Late Movers	[651.7]	[214.6]
Firm Type: Others	−2,146*	−560.2
	[1,137]	[557.0]
Logged MRP	1,179***	1,160***
	[285.8]	[315.4]
Constant	−5,769**	−8,699***
	[2,234]	[2,216]
State Fixed Effects	Yes	No
Developed Status[#]	No	Yes
Molecule Fixed Effects	Yes	Yes
SKU Fixed Effects	Yes	Yes
Year Fixed Effects	Yes	Yes
State × Year Fixed Effects	Yes	No
Developed Status × Year	No	Yes
Units of Observation	Firm–State–Molecule–SKU–Year	Firm–State–Molecule–SKU–Year
Cluster	Firm level	Firm level
Number of Firms	50	50
Number of Molecules	22	22
Number of States	19	19
Number of Years	5	5
Number of SKUs	386	386
Observations	12,636	12,636
R-squared	0.268	0.249

Source: Authors' own calculation using data sourced from AIOCD.

Notes: 1. Robust standard errors are given in brackets.

2. *** $p < 0.01$, ** $p < 0.05$, * $p < 0.1$.

3. [#] The states have been grouped according to the developed status provided in the Rajan Committee Composite Development Index. In case of Delhi, the committee report does not categorize it under any group; however, we have placed it in the relatively developed category. Further, all north-eastern states have been placed in the less-developed category. Within domestic entities, firms which are early movers (the omitted category in these models) have an advantage over everybody else in producing higher quantities.

TABLE 5A.6   Quantity Models—Firm Heterogeneity Based on Headquarters/
Location

OLS Regressions	Dependent Variable: Actual Units Sold	
	Model 1	Model 2
Dummy: Firms in Gujarat	−13.69	−1,897
	[454.9]	[1,445]
Dummy: Firms in Other	3,218***	3,064***
States	[399.2]	[121.8]
Logged Price	1,179***	1,160***
	[288.2]	[317.9]
Constant	−10,186***	−12,975***
	[2,708]	[2,591]
State Fixed Effects	Yes	No
Developed Status#	No	Yes
Molecule Fixed Effects	Yes	Yes
SKU Fixed Effects	Yes	Yes
Year Fixed Effects	Yes	Yes
State × Year Fixed Effects	Yes	No
Developed Status × Year	No	Yes
Units of Observation	Firm–State–Molecule–SKU–Year	Firm–State–Molecule–SKU–Year
Cluster	Firm level	Firm level
Number of Firms	50	50
Number of Molecules	22	22
Number of States	19	19
Number of Year	5	5
Number of SKUs	386	386
Observations	12,488	12,488
R-squared	0.268	0.249

Source: Authors' own calculation using data sourced from AIOCD.

Notes: 1. Robust standard errors are given in brackets.

2. *** $p < 0.01$, ** $p < 0.05$, * $p < 0.1$.

3. Derived by estimating Equation 1 with the heterogeneity classification on the basis of domestic firms' headquarters location.

4. # The states have been grouped according to the developed status provided in the Rajan Committee Composite Development Index. In case of Delhi, the committee report does not categorize it under any group; however, we have placed it in the relatively developed category. Further, all north-eastern states have been placed in the less-developed category.

Empirical Model for Prices

Equation 2 extends our analysis of firm heterogeneity on the ability of firms to charge prices heterogeneously. The AIOCD data provided brand-level prices charged to the retailer and to the end consumer, and these are used to estimate pricing equations, following Danzon and Chao (2000).

$$P_{i, s, m, b, t} = \alpha_0 + \beta_1 (\text{Dummy for Firm Heterogeneity}_i)$$
$$+ \beta_2 (\text{Delivery System}_b) + \alpha_s + \mu_m + \beta_b + \tau_t + \alpha_s \times \tau_t$$
$$+ \varepsilon_{i, s, m, b, t} \tag{2}$$

$P_{i, s, m, b, t}$ measures here a firm i's yearly average price of a focal brand b in a focal molecule, m, for a state, s. Like in Equation 1, to control for unobserved heterogeneity at the state, molecule, brand, and year level, we introduce a cohort of fixed effects (α_s, μ_m, β_b, τ_t, $\alpha_s \times \tau_t$) in Equation 2 as part of our identification strategy. In addition, we control here for the complexity of producing a delivery system for a brand by introducing a dummy delivery system.[38] The 'dummy for firm heterogeneity' is measured in the pricing equations based on whether a firm is a domestic or multinational entity. The sign and significance of β_2 will help us understand how firms are heterogeneously abled to price large molecule brands in the country.

These OLS regressions were estimated using an unbalanced panel from the AIOCD data and using Equation 2 (Table 5A.7). The dependent variable is prices: models 1 and 2 use prices to retailer at the brand/SKU level; and models 3 and 4 use the maximum retail prices. The specifications control for fixed effects at various levels to account for unobserved heterogeneity. Indian firms are associated with lower prices compared to their multinational counterparts.

---

[38] This dummy is 1 for vials, 2 for syringes and injections, and 3 for cartridges, pen-fill, pen-mate, opti-set, flex-pen, and needle (various different forms of new drug delivery systems).

TABLE 5A.7 Pricing Equations—Firm Heterogeneity Based on Domestic/Multinational Firms

OLS Regressions	DV: Price to Retailer		DV: Maximum Retail Price	
	Model 1	Model 2	Model 3	Model 4
Dummy (1= Indian, 0 = Multinational)	−182.8***	−160.8***	−236.9***	−211.1***
	[25.96]	[40.90]	[32.32]	[52.24]
Dummy: Delivery System: Syringe	−138.8***	−148.5***	−149.4***	−157.3***
	[37.65]	[34.64]	[42.02]	[48.73]
Dummy: Delivery System: Others	−186.0***	−219.9***	−225.0***	−264.0***
	[66.19]	[37.59]	[79.85]	[48.71]
Constant	1,905***	2,078***	2,391***	2,598***
	[118.1]	[52.68]	[139.0]	[62.74]
State Fixed Effects	Yes	No	Yes	No
Developed Status#	No	Yes	No	Yes
Molecule Fixed Effects	Yes	Yes	Yes	Yes
SKU Fixed Effects	Yes	Yes	Yes	Yes
Year Fixed Effects	Yes	Yes	Yes	Yes
State × Year Fixed Effects	Yes	No	Yes	No
Developed Status × Year	No	Yes	No	Yes
Units of Observation	Firm–State–Molecule–SKU–Year	Firm–State–Molecule–SKU–Year	Firm–State–Molecule–SKU–Year	Firm–State–Molecule–SKU–Year
Cluster	Firm level	Firm level	Firm level	Firm level

Observations	20,212	20,212	20,212	20,212
R-squared	0.758	0.756	0.756	0.753

Source: Authors' own calculation using data sourced from AIOCD.

Notes: 1. Across all models, the number of firms: 70; number of molecules: 22; number of states: 19; number of years: 5; and number of SKUs: 592.

2. Robust standard errors are given in brackets.

3. *** $p < 0.01$, ** $p < 0.05$, * $p < 0.1$.

Empirical Models for Economies of Scope

A final dimension in our analysis was an exploration of firms' ability to offer product varieties heterogeneously. Equation 3 explores this by specifying how the count of SKUs/brands (with non-zero sales) for a firm i, in a state-region s, for a molecule m, in a year t varies as a function of the firm heterogeneity dummy, controlling for average prices at the retailer level for a firm in a state for a molecule in a year.

$$C_{i, s, m, t} = \alpha_0 + \beta_1(\text{Dummy for Firm Heterogeneity}_i)$$
$$+ \beta_2(\text{Prices}_{i, s, m, t}) + \alpha_s + \mu_m + \tau_t + \alpha_s \times \tau_t + \varepsilon_{i, s, m, t} \qquad (3)$$

Identification is further achieved by introduction of our cohort of fixed effects, as discussed in Equations 1 and 2, to account for various sources of unobserved heterogeneity. The dependent variable is count of brands/SKUs sold by a focal firm in a state-region for a certain molecule in a certain year with non-zero sales. Given the count data nature of the dependent variable, the models were estimated using the Poisson specifications with control for fixed effects at the state–molecule–year level for a firm. In addition, these models were estimated on fewer observations as compared to the more disaggregate firm–state–molecule–brand–year level used to estimate models specified in Equations 1 and 2 (Table 5A.8).

TABLE 5A.8   Firm-level Ability to Launch Product Variety

Poisson Regressions	Model 1: Poisson	Model 2: Poisson	Model 3: Poisson
Firm Type: Followers/	−0.425***	−0.423***	−0.423***
Late movers	[0.118]	[0.113]	[0.113]
Firm Type: Others	−0.404***	−0.396***	−0.397***
	[0.138]	[0.137]	[0.136]
Prices at Retailer Level	0.158***	0.153***	0.153***
(Averaged over Firm–	[0.0460]	[0.0452]	[0.0456]
State–Molecule–Year)			
Constant	0.240	0.205	0.193
	[0.310]	[0.291]	[0.295]

(Cont'd)

TABLE 5A.8 *(Cont'd)*

Poisson Regressions	Model 1: Poisson	Model 2: Poisson	Model 3: Poisson
State Fixed Effects	No	Yes	Yes
Molecule Fixed Effects	Yes	Yes	Yes
Year Fixed Effects	No	Yes	Yes
State × Year Fixed Effects	No	No	Yes
Units of Observation	Firm–State–Molecule–Year	Firm–State–Molecule–Year	Firm–State–Molecule–Year
Cluster	Firm level	Firm level	Firm level
Number of Firms	50	50	50
Number of Molecules	22	22	22
Number of States	19	19	19
Number of Years	5	5	5
Observations	5,271	5,271	5,271
Log Likelihood	−8,522	−8,467	−8,451

Source: Authors' own calculation using data sourced from AIOCD.

Notes: 1. Robust standard errors in parentheses.

2. *** $p < 0.01$, ** $p < 0.05$, * $p < 0.1$.

3. Derived by estimating Equation 3.

References

Agarwal, A. and I. Cockburn. 2003. 'The Anchor Tenant Hypothesis: Exploring the Role of Large, Local, R&D-intensive Firms in Regional Innovation Systems', *International Journal of Industrial Organization*, 21(9): 1227–53.

Ahuja, G. and S. Yayavaram. 2011. 'Explaining Influence Rents: The Case for an Institutions-based View of Strategy', *Organization Science*, 22(6): 1631–52.

Alliance for Safe Biologic Medicines. 2012. 'ASBM Survey Reveals Key Safety Concern Associated with Biosimilar Naming; Urges Distinct Naming Requirement', Press release, 17 September. Available at http://safebiologics.org/resources/2012/09/asbm-survey-reveals-key-safety-concern-associated-with-biosimilar-naming-urges-distinct-naming-requirement/, last accessed on 29 March 2017.

Arora, A., A. Fosfuri, and A. Gambardella. 2001. 'Markets for Technology and their Implications for Corporate Strategy', *Industrial and Corporate Change*, 10(2): 419–51.

Arora, A. and A. Gambardella. 1994. 'The Changing Technology of Technological Change: General and Abstract Knowledge and the Division of Innovative Labour', *Research Policy*, 23(5): 523–32.

Azoulay, P., J. Graff Zivin, and B. Sampat. 2012. 'The Diffusion of Scientific Knowledge across Time and Space: Evidence from Professional Transitions for the Superstars of Medicine'. Available at http://pazoulay.scripts.mit.edu/docs/diffusion_rdie.pdf, last accessed on 29 March 2017.

Babu, G. 2012. 'IPAB Sets aside Indian Patent for Roche's Pegasys', *Business Standard*, 2 November. Available at http://www.business-standard.com/article/companies/ipab-sets-aside-indian-patent-for-roche-s-pegasys-112110200166_1.html, last accessed on 29 March 2017.

Bajaj, V. and A. Pollack. 2012. 'India's Supreme Court to Hear Dispute on Drug Patents', *The New York Times*, 6 March. Available at http://www.nytimes.com/2012/03/07/business/global/indias-supreme-court-to-hear-long-simmering-dispute-on-drug-patents.html?_r=0, last accessed on 29 March 2017.

Branstetter, L., C. Chatterjee, and M.J. Higgins. 2014. 'Starving (or Fattening) the Golden Goose? Generic Entry and the Incentives for Early-stage Pharmaceutical Innovation', NBER Working Paper 20532.

Bresnahan, T.F. and M. Trajtenberg. 1995. 'General-purpose Technologies: "Engines of Growth"?' *Journal of Econometrics*, 65(1): 83–108.

Campaign for Access to Essential Medicines. 2011. 'Untangling the Web of Antiretroviral Price Reductions'. Available at http://apps.who.int/medicinedocs/documents/s18716en/s18716en.pdf, last accessed on 29 March 2017.

Chatterjee, C. 2011. 'Intellectual Property, Incentives for Innovation and Welfare: Evidence from the Global Pharmaceutical Industry', Doctoral dissertation, Carnegie Mellon University, Pittsburgh, PA.

Chatterjee, C., K. Kubo, and V. Pingali. 2015. 'The Consumer Welfare Implications of Governmental Policies and Firm Strategy in Markets for Medicines', *Journal of Health Economics*, 44: 255–73.

Chaudhuri, S. 2005. *The WTO and India's Pharmaceuticals Industry: Patent Protection, TRIPS, and Developing Countries*. New Delhi: Oxford University Press.

Cohen, S.N., A.C. Chang, H.W. Boyer, and R.B. Helling. 1973. 'Construction of Biologically Functional Bacterial Plasmids In Vitro', *Proceedings of the National Academy of Sciences*, 70(11): 3240–4.

Danzon, P.M. and L.W. Chao. 2000. 'Cross-national Price Differences for Pharmaceuticals: How Large, and Why?', *Journal of Health Economics*, 19(2): 159–95.

Generics and Biosimilars Initiative Journal. 2013. '"Similar Biologics" Approved and Marketed in India', *Generics and Biosimilars Initiative Journal*, 2(1): 50–1. Available

at http://gabi-journal.net/similar-biologics-approved-and-marketed-in-india.
html, last accessed on 29 March 2017.

Grabowski, H., I. Cockburn, and G. Long. 2006. 'The Market for Follow-on Biologics: How will it Evolve?', *Health Affairs*, 25(5): 1291–301.

Henderson, R. and I. Cockburn. 1994. 'Measuring Competence? Exploring Firm Effects in Pharmaceutical Research', *Strategic Management Journal*, 15(S1): 63–84.

————. 1996. 'Scale, Scope, and Spillovers: The Determinants of Research Productivity in Drug Discovery', *RAND Journal of Economics*, 27(1): 32–59.

India Infoline. n.d. 'Dr. Reddy's and Merck Serono to develop and commercialise Biosimilars'. Available at http://www.indiainfoline.com/article/news-top-story/dr-reddy-s-and-merck-serono-to-develop-and-commercialise-biosimilars-113103107899_1.html, last accessed on 29 March 2017.

Kapczynski, A. 2013. 'Engineered in India—Patent Law 2.0', *New England Journal of Medicine*, 369 (6): 497–9.

Kapoor, R. and T. Klueter. 2013. 'Decoding the Adaptability–Rigidity Puzzle: Evidence from Biopharmaceutical Incumbents' Pursuit of Gene Therapy and Monoclonal Antibodies', Working paper. Available at https://agenda.unibocconi.it/eventi/attach/Kapoor%20Paper20131007125109.pdf, last accessed on 29 March 2017.

Kastelle, T. 2013. 'Pasteur's Quadrant'. Available at http://timkastelle.org/blog/2013/05/innovation-requires-a-bias-towards-action/pasteurs-quadrant/#lightbox/0/, last accessed on 29 March 2017.

Kulkarni, Kaustubh and Henry Foy. 2012. 'Pfizer Scraps Insulin Deal with India's Biocon', Reuters, 13 March. Available at http://in.reuters.com/article/us-pfizer-biocon-idUSBRE82C05920120313, last accessed on 29 March 2017.

McGee, J. and H. Thomas. 1986. 'Strategic Groups: Theory, Research and Taxonomy', *Strategic Management Journal*, 7(2): 141–60.

Moxley, S. 2012. 'Designer Drugs: The Quest for a Rational Therapeutics in the Biotechnology Era, 1973–97', Doctoral dissertation, Carnegie Mellon University, Pittsburgh, PA.

Oettl, A. 2012. 'Reconceptualizing Stars: Scientist Helpfulness and Peer Performance', *Management Science*, 58(6): 1122–40.

Patel, P. and K. Pavitt. 1997. 'The Technological Competencies of the World's Largest Firms: Complex and Path-dependent, but Not Much Variety', *Research Policy*, 26(2): 141–56.

Pauling, L., H.A. Itano, S.J. Singer, and I.C. Wells. 1949. 'Sickle-cell Anemia: A Molecular Disease', *Science*, 110(2865): 543–48.

Pfizer. 2010. 'Biocon and Pfizer Enter Into Global Commercialization Agreement: Creates Global Alliance Well Positioned to Deliver Essential Insulin Treatments to Diabetes Patients Worldwide', Press Release, 18 October. Available at http://press.pfizer.com/press-release/biocon-and-pfizer-enter-global-commercialization-agreement, last accessed on 29 March 2017.

Roach, M. and H. Sauermann. 2010. 'A Taste for Science? PhD Scientists' Academic Orientation and Self-selection into Research Careers in Industry', *Research Policy*, 39(3): 422–34.

Shanker, Archana. 2016. 'The battle for biotech drugs in India', *Life Sciences Intellectual Property Review*, 9 June. Available at http://www.lifesciencesipreview.com/contributed-article/the-battle-for-biotech-drugs-in-india, last accessed on 29 March 2017.

Shaver, J.M. and F. Flyer. 2000. 'Agglomeration Economies, Firm Heterogeneity, and Foreign Direct Investment in the United States', *Strategic Management Journal*, 21(12): 1175–93.

Sidhartha. 2013. 'India Counters US over Patents', *The Times of India*, 25 June, available at http://timesofindia.indiatimes.com/business/india-business/India-counters-US-over-patents/articleshow/20753719.cms, last accessed on 29 March 2017.

Stokes, D.E. 1997. *Pasteur's Quadrant: Basic Science and Technological Innovation.* Washington, DC: Brookings Institution Press.

Teece, D.J. 1986. 'Profiting from Technological Innovation: Implications for Integration, Collaboration, Licensing and Public Policy', *Research Policy*, 15(6): 285–305.

The Hindu. 2010. 'Ranbaxy Signs Pact with US Firm for Developing Biosimilars', 29 March. Available at http://www.thehindu.com/business/Ranbaxy-signs-pact-with-US-firm-for-developing-biosimilars/article16627708.ece, last accessed on 29 March 2017.

Watson, J.D. and H.C.F. Crick. 1953. 'A Structure for Deoxyribose Nucleic Acid', *Nature*, 171(4356): 737–38.

6

SHYAMKRISHNA BALGANESH
DAVID NIMMER

Fair Use and Fair Dealing

Two Approaches to Limitations and Exceptions in Copyright Law

Copyright is predicated on realizing a balance between affording protection to creators of original expression and granting access to such expression to members of the public. Copyright law's framework of exclusive rights represents its form of protection, whereas the various 'limitations and exceptions' that the law recognizes to these rights constitute its mechanism of enabling access. Limitations and exceptions, thus being central to copyright's basic analytical framework, are hardly orthogonal to the system.

While copyright's grant of exclusive rights contributes to a country's innovation policy by providing the actors with a market-based incentive for creativity, its limitations and exceptions are equally central to that policy. First, they enable the public to access and use protected expression when doing so is determined to be socially beneficial (for example, educational uses). In addition, and just as importantly, they enable additional creativity by allowing users and consumers to use and copy protected expression during the process of producing new expression themselves.

Countries around the world adopt different approaches in structuring these limitations and exceptions into their copyright laws. Broadly speaking, these approaches take two forms: the 'fair use approach' and the 'fair dealing approach'. The former originated in the copyright law of the United States (US), deriving from judge-made law requiring courts to consider a variety of contextual factors before deciding whether a defendant's use/copying of a protected work qualifies as non-infringing. It relies heavily on incremental decision-making, with the law developing ex post. The latter approach, on the other hand, is followed in the copyright laws of most commonwealth countries and continental jurisdictions. Here, the copyright statute generally delineates specific types of uses and forms of copying that are categorically exempted from the scope of copyright infringement, with no need for further contextual inquiry once the use is found to fit into a particular box or category. The fair dealing approach relies much less on courts than the fair use approach, but necessitates extensive legislative tailoring of the copyright statute to ensure that a wide variety of uses are exempted over time, as conditions become varied and technology progresses.

This chapter undertakes a comparative analysis of these two approaches to structuring copyright limitations and exceptions, using the copyright law of India as the principal point of focus and comparing it to the approach adopted in the US. As traditionally conceived, the fair use approach is thought to allow for greater flexibility and adaptability in the creation of limitations and exceptions, while the fair dealing approach—being more rigid—is believed to produce greater certainty in the law. Although this dichotomy is routinely emphasized in discussions of the two approaches, we argue that a comparative study of the Indian and American approaches suggests that the divergence is far from obvious. Despite Indian copyright law's adherence to the fair dealing approach, Indian courts are reluctant to cede decision-making completely to the legislature and continue to play an important role in defining copyright's various limitations and exceptions. Indeed, at times, they even introduce new ones not present in the statute. Over time, Indian courts have interpreted the statute's fair dealing provisions and applied them in terms largely similar to the fair use doctrine in the US. Conversely, in the US, the flexibility of fair use has, over time, produced rigid categories of exceptions that seem to operate much like bright-line rules, rather than open-ended standards. We thus conclude that, in discussing the two approaches, copyright scholars and lawyers should be wary of broad generalizations, which depend on a variety of institutional factors that are largely exogenous to the domain of copyright law.

The first section of this chapter begins with a brief overview of the two approaches, using Indian and the US copyright law as exemplars. It sets out the broad ideas underlying each and identifies the primary analytical variables at stake. The second section examines whether the dichotomy traditionally understood between the two regimes is indeed as watertight as it is claimed to be; it suggests that copyright scholars adopt a more nuanced and contextual approach to this dichotomy, looking to the actual working of the copyright system in each instance and the institutional variables involved that tend to influence the regime's approach to limitations and exceptions. To illustrate this point, it draws on examples from both jurisdictions.

Two Models of Copyright Limitations and Exceptions

Limitations and exceptions form an integral part of any copyright system by ensuring that the copying of protected works under certain circumstances and conditions be exempted from the ambit of copyright infringement. Because copyright law is, for the most part and in most countries, statutory in origin, limitations and exceptions embedded within its working generally originate in a statutory directive. Although that feature did not apply to the US copyright law prior to the most recent copyright statute (the Copyright Act of 1976), as of today it remains true of the US no less than of almost all other copyright systems around the world. The domain in which most systems differ is in their process of determining the circumstances and conditions under which copying ought to be exempted.

The two principal models seen around the world can be broadly characterized, as introduced earlier, as the fair use approach and the fair dealing approach. In general terms, the fair use approach involves an ex post, case-by-case method of determining whether a particular instance of copying should be exempted from infringement, usually based on a set of open-ended standards. The fair dealing approach, on the other hand, normally entails a statute delineating specific circumstances and settings in which particular acts of copying are exempted from infringement. Unlike fair use, fair dealing provisions are normally encapsulated by bright-line rules (rather than standards) and tersely worded language. This section elaborates on these two models, using the US and India (built on the United Kingdom [UK] statute) as examples. It describes the two approaches, then details the principal institutional and structural trade-offs that the choice between the two is ordinarily thought to entail.

The Fair Use Approach: The US

Copyright protection in the US dates back to 1790—it was enacted by the first Congress to convene after adoption of the US Constitution. At that early juncture, there was no such thing as a recognized defence. But their roots did not take long to sprout. Justice Joseph Story, considering an 1841 case involving the collected papers of President George Washington, began his remarks by commenting that the case presented 'one of those intricate and embarrassing questions, arising in the administration of civil justice, in which it is not, from the peculiar nature and character of the controversy, easy to arrive at any satisfactory conclusion, or to lay down any general principles applicable to all cases' (*Folsom* v. *Marsh* 1841: 44). Acknowledging that 'the lines approach very near to each other, and, sometimes, become almost evanescent, or melt into each other' (1841: 44), he articulated what would later be codified as the doctrine of 'fair use'. As later summarized, that doctrine 'permits courts to avoid rigid application of the copyright statute when, on occasion, it would stifle the very creativity which that law is designed to foster' (*Stewart* v. *Abend* 1990: 236). Over the intervening decades, courts have followed suit by vindicating the defence of fair use on many occasions and rejecting it in other instances.

A ready measure of the confusion in the courts surrounding fair use can be gleaned by looking at how the US Supreme Court dealt with the issue. Under the 1909 Act that governed for most of the twentieth century, the Supreme Court heard arguments about two fair use cases.[1] In both instances, the court was evenly divided, and hence unable to resolve the dispute before it. No other rulings in the copyright sphere during those decades produced that result of affirmances by an equally divided court.

Over a century later, the Congress decided to grapple with the field. As part of the process of copyright statutory revision, it commissioned a series of reports, one of which was devoted to the fair use cases that had accumulated over more than a century since Justice Story had handed down his 1841 ruling. After canvassing those opinions, Alan Latman's report (1958) outlined the broad options that the Congress could adopt: (a) maintain statutory silence and allow courts to continue to develop the field; (b) recognize fair use in broad strokes through a statutory provision, without attempting to clarify its application; (c) specify general criteria for how it should apply; or (d) cover specific situations, such as criticism and review.

In 1976, the Congress finally enacted the product of the review that it had initiated two decades earlier. The resulting Copyright Act of 1976 took effect on 1 January 1978, and continues to govern until today. In terms of fair use, the Congress chose a combination of the final two possibilities, by enacting section 107 of the act. That provision begins with a preamble of presumptively fair uses, then sets forth four non-exclusive factors for courts to consider when evaluating fair use:

> The fair use of a copyrighted work, including such use by reproduction in copies or phonorecords or by any other means specified by that section, for purposes such as criticism, comment, news reporting, teaching (including multiple copies for classroom use), scholarship, or research, is not an infringement of copyright. In determining whether the use made of a work in any particular case is a fair use the factors to be considered shall include—
>
> (1) the purpose and character of the use, including whether such use is of a commercial nature or is for non-profit educational purposes;
> (2) the nature of the copyrighted work;
> (3) the amount and substantiality of the portion used in relation to the copyrighted work as a whole; and

---

[1] *Columbia Broadcasting System, Inc. v. Loew's Inc.* (1958); *Williams & Wilkins Co. v. United States* (1975).

(4) the effect of the use upon the potential market for or value of the copyrighted work.

Here, a word is in order specifying the relationship between the preamble and the factors that follow. First, if the utilization qualifies thereunder—for instance, as criticism or news reporting—the fair use case is bolstered. Nonetheless, it does not become presumptively fair on that basis alone and must still be evaluated in terms of the four factors that follow. At times, these considerations can push in opposite directions. *The New York Times*, for instance, is a commercial enterprise that sells copies for its own profit. Its status as a news organization favours fair use, whereas its commercial nature might incline the first factor against fair use. Obviously, a more searching inquiry is necessary.

One of the matters specified in the preamble is 'teaching (including multiple copies for classroom use)'. Along those lines, the legislative history accompanying the current act—a report from the House of Representatives (1976)—contained elaborate guidelines addressing how many copies could be made for that purpose, encompassing what percentage of a work of authorship is so produced, in the context of how spontaneous or planned the educational purpose was, and likewise setting forth other criteria. Inasmuch as those guidelines were not themselves passed into law, their footing has remained uncertain ever since.

Copyright litigation in the US is a legion; a host of cases have arisen to interpret its every feature. In the fair use domain, that litigation has been particularly pronounced. In fact, the Supreme Court has decided more copyright cases posing the matter of fair use under the 1976 Act than any other aspect of copyright law. It is a measure of the dissension in the field that practically every such case arose in the posture of a district court holding reversed by the court of appeals and, in turn, reversed by the Supreme Court, usually with a dissenting opinion at the highest level.

We have already called out the preambular language of 'teaching (including multiple copies for classroom use)' and the House of Representatives report guidelines for its implementation. It remains to be added that litigation results regarding photocopying for research and teaching purposes have been spotty. One of the pre-1976 cases resulting in an affirmance by an equally divided Supreme Court arose precisely in that context (*Williams & Wilkins Co.* v. *United States* 1975). The result was to validate photocopying of scholarly journals undertaken by the National Institutes of Health (NIH). But the opposite result emerged from a different circuit court some years into the pendency of the current Act. In 1994, the Second Circuit ruled that fair use did not protect making copies of articles out of the *Journal of Catalysis* and similar scholarly materials for the purpose of aiding a private corporation's scientific research. A significant intervening development was the advent of the Copyright Clearance Center, which offered affordable licences for such material as the *Journal of Catalysis* that had not been available to the NIH decades earlier. Two years later, the Sixth Circuit likewise ruled against a service producing 'course packs' at the behest of university professors. Those course packs consisted largely of excerpts from journals and books that were relevant to a single subject being taught in a university course; their designation 'for classroom use' failed to win favourable treatment. Nonetheless, when the wheel turned once again, the opposite result inured. Cambridge University Press challenged Georgia State University's 2009 copyright policy, which allowed electronic copies to be made of individual chapters from various books for student use. The district court initially validated the 'course packs', ruling them to be protected as fair use (*Cambridge Univ. Press* v. *Becker* 2012). On appeal however, the Eleventh Circuit reversed the district court's opinion, finding that such use by the university did not amount to fair use (*Cambridge Univ. Press* v. *Patton* 2014). It should be added that the Copyright Clearance Center granted licences for about 60 per cent of the works in Cambridge University Press's catalogue—but only print copies, not electronic copies. This thumbnail sketch of divergent results in one discrete field suffices to highlight the shifting sands over time in fair use determinations.

The other pre-1976 case resulting in an affirmance by an equally divided Supreme Court arose in the context of parody (*Columbia Broadcasting System, Inc.* v. *Loew's Inc.* 1958). Famous comedian Jack Benny produced a half-hour television (TV) show in 1952, burlesquing the motion picture *Gaslight*. When sued, he defended on the grounds of parody. The Ninth Circuit denied that defence, as it later denied the fair use defence offered by a comic book portraying Mickey Mouse and other Disney characters 'as active members of a free thinking, promiscuous, drug ingesting counterculture' (*Walt Disney Productions* v. *Air Pirates* 1978: 753). Subsequently, the Supreme Court validated the parody defence proffered by a rap group that replicated much of the music and lyrics from the country song 'Pretty Woman' (*Campbell* v. *Acuff-Rose Music, Inc.* 1994). In that context, it highlighted the difference between protected 'parody' and less protected 'satire'—the former comments on the original author's work, whereas the latter 'has no critical bearing on the substance or style of the original composition' (*Campbell* v. *Acuff-Rose Music, Inc.* 1994: 580). Of course, that line itself is often difficult to draw.

Each of the four factors set forth in the statute has a tendency to turn into a bright-line rule. Thus, at one point, utilizations that were commercial tended to be disfavoured under the first factor, and to routinely lead to a denial of fair use. Copying from unpublished works tended to be disfavoured under the second factor, and to routinely lead to a denial of fair use. Copying the totality of a work works tended to be disfavoured under the third factor, and to routinely lead to a denial of fair use. Depriving the copyright owner of income it would have otherwise earned tended to be disfavoured under the fourth factor, and to routinely lead to a denial of fair use.

Both the Congress and the Supreme Court have tried to halt that tendency in its tracks. In terms of the solicitude shown to unpublished works under the second factor, a series of court rulings threatened to hold all quotations from unpublished manuscripts categorically unfair. The matter got so dire that the Congress intervened in 1992. In the only amendment to the wording of the statutory provision (17 U.S.C.A. §107 [1992]), it added the following qualification at the end: 'The fact that a work is unpublished shall not itself bar a finding of fair use if such finding is made upon consideration of all the above factors.' In a like measure, a unanimous Supreme Court case (*Campbell* v. *Acuff-Rose Music, Inc.* 1994) commanded that all factors be taken into account, rejecting any bright-line rule derived from a single factor.

On account of these interventions, commercial utilizations today stand a chance to be characterized as fair use. Under the first factor, for instance, an artist who produced appropriation art by engaging in wholesale copying from the products of the plaintiff photographer, which the defendant in turn sold for millions of dollars, prevailed on his fair use defence (*Blanch* v. *Koons* 2006). Under the second factor, a service that copied unpublished student papers to test them on behalf of schools through a 'plagiarism detection service' likewise prevailed in its fair use defence (*A.V. ex rel. Vanderhye* v. *iParadigms, LLC* 2009). Under the third factor, a group of university libraries that copied whole books in their collections proved similarly victorious (*Authors Guild, Inc.* v. *HathiTrust* 2014). Further, under the fourth factor, *Playboy* magazine prevailed when it reproduced an old high-school photograph of one of its Playmates of the Month without paying the photographic service the fee it typically charged for all reproductions (*Carla Calkins* v. *Playboy Enterprises Intern., Inc.* 2008).

Integral to each of the cases just canvassed is that the use in question qualified as 'transformative'. Deriving from a scholarly article by Judge Pierre Leval (1990) that was later adopted by the Supreme Court, this aspect enquires into whether 'the new work merely supersedes the objects' of the original creation or instead adds something new, with a further purpose or different character, altering the first with new expression, meaning, or message. We return to the transformative test later. Unfortunately, its application has been so broad as to qualify it as almost all things to all people.

Having articulated four factors, how does the statute weigh each? In other words, what if factors 1 and 3 favour fair use, but factors 2 and 4 are to the contrary? What if factor 4 strongly favours fair use, factors 1 and 2 are weakly to the contrary, and factor 3 is neutral? Scores of permutations are cognizable.

Unfortunately, the statute is silent as to relative weight. Indeed, the statute itself does not foreclose the application of additional factors, unenumerated in the congressional language. Courts must therefore consider each of the four factors, along with additional circumstances, and apply their own calculus as to the ultimate resolution. Cases are decided ad hoc, with no certainty in advance how a given case will pan out.

The Fair Dealing Approach: India

India's current copyright law is contained in the Copyright Act of 1957. This was the country's first post-Independence copyright legislation and replaced the Copyright Act of 1914, which was enacted by the

British colonial government for India. The Copyright Act of 1957 drew extensively from its UK counterpart, the UK Copyright Act of 1956. Since the UK Act adopted a fair dealing approach to limitations and exceptions, the Indian statute followed suit in large measure. Sections 6 through 10 of the UK Act dealt with that statute's various limitations and exceptions. While some provisions consciously used the phrase 'fair dealing', others merely described particular instances of copying with some level of detail and went on to exempt those acts from infringement altogether.

The Indian Copyright Act of 1957 consolidates its various limitations and exceptions in one principal section of the statute, section 52. Dealing with actions that 'do not constitute copyright infringement', this section contains what are in essence two principal kinds of exceptions to infringement. The first kind involves what are true fair dealing provisions, which qualify their description of the activity that is exempted by the phrase 'fair dealing', thereby suggesting that the activity described is entitled to an exemption only when assessed to be a fair dealing under the law, not simply when found to have been undertaken as a factual matter. The second kind does just the opposite. These latter provisions merely describe an activity in some detail, and allow the exception to be invoked upon a mere showing that the factual elements identified in the provision (in its description of the activity) have been met. Although these latter types of exceptions, strictly speaking, do not qualify as fair dealing, inasmuch as they are interspersed amongst the true fair dealing ones, the rules of construction and interpretation that courts apply to one generally carry over to the other as well. For the purposes of our discussion, we treat them both as falling within the fair dealing approach, and in situations where the precise construction of the statute makes a difference, we draw out that aspect.

Consider the following examples, illustrative of the two kinds of exceptions just described. Section 52(b) exempts from infringement the following action:

[A] fair dealing with a literary, dramatic, musical or artistic work for the purpose of reporting current events—

   (i) in a newspaper, magazine or similar periodical, or

   (ii) by broadcast or in a cinematograph film or by means of photographs.

As should be obvious, the provision specifies a particular kind of copying that qualifies for the exemption: copying of a work for news reporting. It further specifies the factual contours of the activity that meets the requirements, namely that the news reporting must be of 'current events' and be in a periodical or in a broadcast, film, or photograph. Yet, as a preliminary, it requires the court—or other decision-maker—to be assured that the copying in question was not just any copying but a 'fair dealing'. This raises the obvious question of what exactly a fair dealing is, and how a court is to go about determining the proposition. Sadly, the statute's definition section provides no guidance whatsoever on that score.

Since the Act of 1957 drew extensively—in structure, design, and substance—from its UK counterpart, Indian courts and lawmakers quite naturally looked to English courts to figure out what exactly fair dealing meant. One of the leading cases on fair dealing in the UK sought to define and lay down a formula for courts to apply in giving the words 'fair dealing' operational content in individual cases. *Hubbard* v. *Vosper* (1972) involved an interpretation of section 6 of the UK Act in an interesting posture. The Church of Scientology had commenced an action for copyright infringement (through its founder, L. Ron Hubbard) against a former member who had published a book criticizing the organization. The plaintiff alleged that the defendant's book contained material that had been copied from the plaintiff's own books and writing. The defendant raised the defence of fair dealing and succeeded. Writing for the Court of Appeal, Lord Denning—a legendary Law Lord—took the opportunity to clarify the scope of the fair dealing defence and its meaning:

> It is impossible to define what is 'fair dealing'. It must be a question of degree. You must consider first the number and extent of the quotations and extracts. Are they altogether too many and too long to be fair? Then you must consider the use made of them. If they are used as a basis for comment, criticism or review, that may be fair dealing. If they are used to convey the same information as the author, for a rival purpose, that may be unfair. Next, you must consider the proportions. To take long extracts and attach short comments may be unfair. But, short extracts and long comments may be fair. Other considerations may come to mind also. But, after all is said and done, it must be a matter of impression. As

with fair comment in the law of libel, so with fair dealing in the law of copyright. The tribunal of fact must decide. In the present case, there is material on which the tribunal of fact could find this to be fair dealing. (*Hubbard* v. *Vosper* 1972: 94)

Lord Denning's observations have since been taken to represent the approach that courts are to use in determining when a defendant's copying is fair, even when the other statutorily delineated factual elements have been satisfied. Indian courts have, for the most part, taken *Hubbard*'s observations to represent the principal way of interpreting and applying fair dealing even under the act of 1957. Notwithstanding that subsequent UK courts have sought to cabin Lord Denning's observations, Indian high courts continue to treat his elements and factors as the principal bases with which to approach section 52's fair dealing provisions. Indeed, on occasion, they have expanded on his insights and added additional elements.

Outside of the true fair dealing provisions, section 52 also contains exceptions which, as noted earlier, specify certain kinds of copying and designate them as non-infringing and as a factual matter. These provisions require little subjective judgement on the part of courts, as to their fairness or otherwise. Section 52(c) furnishes a good example. It provides that 'the reproduction of a literary, dramatic, musical or artistic work for the purpose of a judicial proceeding or for the purpose of a report of a judicial proceeding' is not an infringement. In these kinds of provisions, a defendant must merely establish (and the court must be satisfied) that the precise contours of the statutory exception have been satisfied: once this showing is met, the activity is treated as non-infringing. No further balancing or analysis is required on the part of the court.

Although the first category of exceptions may seem markedly different from the second insofar as they delegate the determination of fairness to courts, both categories together exhibit an important common feature that typifies the fair dealing approach as a whole: courts are meant to interpret the circumstances and conditions mandated by the statutory exception with a heightened level of rigidity. To take our examples from the Indian statute, courts are to thus interpret the contours of 'a periodical' and what constitutes a 'judicial proceeding' in equally strict and narrow terms. In referring to this fundamental process of construction—applicable to all fair dealing exceptions—one English court thus characterized the fair dealing approach as accepting that the provisions 'define with extraordinary precision and rigidity the ambit of various exceptions to copyright protection' (*Pro Sieben Media AG* v. *Carlton UK Television Ltd* 1997: 516). Courts have very little leeway in expanding these provisions based on what might be deemed independently reasonable under the circumstances.

A recent decision of the Delhi High Court vividly illustrates the facial rigidity of the fair dealing approach. Applying the exemption provisions of the Indian Copyright Act, *Super Cassettes Indus. Ltd.* v. *Chintamani Rao* (2012) confronted whether subsections 52(1)(a) and (b) could immunize the defendant's actions from liability, insofar as those activities related to the plaintiff's protected cinematographic films and sound recordings. Those two subsections define fair dealing to apply to specified utilization of a 'literary, dramatic, musical or artistic work'. Neither makes mention of a cinematographic film or a sound recording, both of which, the defendant's lawyer argued, are in a sense derivative of traditional 'literary, artistic, dramatic, and musical works', concluding on that basis that section 52 should be liberally and purposively construed to embrace these additional categories of (not specifically enumerated) works. The court rejected the defendant's argument, and in so doing elaborated on how it conceptualized its task in interpreting the act's various exceptions and limitations:

> Section 52 carefully and exhaustively enlists various actions which would not constitute infringement of copyright in different classes of works and the limits on such use…. [The] exceptions in Section 52 are carefully crafted and are use-specific as well as work-specific. Each clause makes clear both—the type and class of work to which it applies, and the particular exempted use of such work. (*Super Cassettes Indus. Ltd.* v. *Chintamani Rao* 2012: 22)

The court thus interpreted section 52 to be exhaustive in structure, with little room for further elaboration or expansion by courts independent of legislative authorization. It went on to conclude that 'Parliament deliberately and consciously chose the class of works in relation to which it permitted the exploitation of the copyright for specific purposes only'

(*Super Cassettes Indus. Ltd.* v. *Chintamani Rao* 2012: 22). In addition to interpreting the rigidity of the provision to be a conscious effort to limit the applicability of the exceptions, the court also sought to justify it substantively, as embodying the legislature's calculated wisdom relating to the types of works in question: 'There is very good reason for not including cinematograph films and sound recordings in clauses (a) and (b) in Section 52(1) of the Act. Being derived works, cinematograph films and sound recordings involve much greater financial investment when compared to investment that may have been made in the creation of original literary, dramatic, musical and artistic work' (*Super Cassettes Indus. Ltd.* v. *Chintamani Rao* 2012: 23).

The court's reasoning in the case aptly illustrates everything that the fair dealing approach entails, at least in theory: (a) a narrow, textualist, and rigid reading of the wording in each statutory exception, accompanied by no effort to exercise judicial discretion or supplementary lawmaking on a delegated basis; (b) a belief that the power for all lawmaking in relation to such exceptions and limitations inheres in the legislative branch, with the courts doing no more than applying the law as articulated by the legislature; (c) exercising discretion only when expressly delegated, with the obvious area of delegation being the determination of fairness in relation to the copying; and (d) the tendency to treat the legislative guidance embodied in the statute as embodying important trade-offs and choices that ought to be followed and discerned, even when not obvious or apparent. *Super Cassettes* thus epitomizes the fair dealing approach which, as noted previously, owes its legacy to the rules formulated by the English common law. As it turns out, however, a subsequent appeal overruled this decision (*India TV Independent News Service Pvt. Ltd. & Ors.* v. *Yashraj Films Pvt. Ltd.* 2013). The appellate on *Super Cassettes* decision abandoned the fair dealing approach altogether, an issue that the next part addresses. For now, what is essential to note is that the Delhi high court's opinion in this matter epitomizes the fair dealing approach to which Indian courts were traditionally accustomed, drawing on English fair dealing cases.

Under the fair dealing approach, then, primary responsibility for limitations and exceptions vests with the legislature. At least in theory, any updating of copyright law in relation to new technologies or means of copying and exploitation must come about through amendment of the statute's text. The power of courts is limited to determining when a given exploitation qualifies as 'fair'; yet, even this power arises solely when the statutory exception in question expressly sets forth the phrase 'fair dealing' to qualify its application.

Trade-offs

Having seen what the two different approaches entail, as illustrated by examples of cases decided in India and the US, it remains to set forth the principal structural and substantive trade-offs that the choice of one model/approach over the other is thought to entail. Three important trade-offs are identified and discussed below.

Guidance

Of the two approaches, the fair use model is often criticized for its inability to provide actors with sufficient guidance about the extent to which their actions are protected against copyright infringement in advance. Since fair use entails a standards-based case-by-case analysis, actors are seen to be dependent on having courts adjudicate their claims in order to obtain the requisite certainty. Given the costs of litigation, the widespread belief is that individuals are encouraged to settle their claims, even if they are legally well grounded; and they are equally incentivized to obtain needless licences, even as to uses that should qualify as fair use. Fair dealing, on the other hand, is thought to do just the opposite. By delineating the scope and extent of the exception with sufficient certainty upfront, users are believed to be in a position to plan their activities in advance, without having to worry about copyright infringement. In short, fair dealing is believed to provide users of copyrighted works (that is, potential defendants) with better guidance than fair use.

Flexibility

Fair use is considered more flexible due to the open-ended character of the four-factor assessment, while the fair dealing approach can potentially become rigid in its application. Flexibility brings along with itself reduced certainty, which in turn has its own costs and benefits. While fair use is believed to provide actors with insufficient guidance owing to its uncertainty, this reduced certainty allows the approach to adapt better to new circumstances and contexts. The rigidity of fair dealing allows actors to determine the scope and extent of different exceptions with greater clarity. Yet, this rigidity impedes (indeed, disallows) courts from adapting these exceptions to new contexts and areas incrementally, as technology and circumstances demand. Fair use, on the other hand, has a clear benefit in this regard, allowing copyright law's limitations and exceptions to develop to meet the needs of the times more expeditiously. Fair use is thus thought to be significantly more flexible and adaptable than fair dealing. Indeed, some have argued that it is the flexibility and open-endedness of fair use that has enabled various fair-use-dependent industries to thrive and flourish in the US.

Lawmaking Institution

The third important trade-off between the two approaches is institutional. It relates to the institution that is recognized to have primary authority in creating exceptions and limitations to copyright's exclusive rights and expanding or extending them to new scenarios. The fair dealing approach emerges from the understanding that the legislature retains primary authority over all of copyright law, including the structure and content of limitations and exceptions, and that courts are to do no more than apply these exceptions to individual cases. Fair use, on the other hand, is conceptualized as a form of deputized lawmaking, wherein the US copyright statute delegates to courts the task of applying the four general principles (enumerated in the statute) to new situations and contexts and, in the process, making new law for those contexts and areas. Unlike fair dealing, fair use does not begin with the premise of legislative supremacy for the creation of copyright limitations and exceptions.

The Distinction in Practice

The previous discussion of the two approaches to copyright exceptions and the various structural trade-offs that they entail would lead one to predict that, in practice, several differences are likely to emerge between the copyright systems choosing one over the other. In systems adopting the fair dealing approach, one might predict that courts would tend to follow the letter of the law strictly, thereby causing the law to lag behind technological and socio-economic changes that have transpired since the statute was most recently amended. This discussion would also suggest that copyright limitations and exceptions remain a closed set in this system, with courts actively cabining their own creativity and innovativeness in crafting new exceptions or in extending the logic of pre-existing exceptions to new situations. Correlatively, as to systems adopting the fair use approach, it predicts that they would be more creative in crafting new exceptions to new situations, and allow the law to keep up more rapidly with changes in technology. In so doing however, the law would compromise its certainty to a great degree, providing minimal guidance to actors, who would have to wait on a case-by-case adjudication before knowing if their acts of copying were, in fact, exempted from infringement.

In practice, however, as we show in the following paragraphs, these predictions prove to be grossly exaggerated in the two systems we compare. Despite India's adoption of a fair dealing approach, its courts have proven to be fairly innovative in interpreting the text of the statute's exceptions and in crafting new exceptions. At the same time, the US courts have strived to inject a good degree of predictability into the working of the fair use doctrine and, over time, have developed discernible patterns that provide actors with a fair measure of guidance in given, concrete situations.

Creativity despite Constraints: The Indian Example

Despite what the fair dealing approach—in theory—suggests that courts should be doing when interpreting and applying the statutorily enumerated exceptions, Indian high courts have been unable to stifle their creative impulses altogether. The opening created by the statute's use of the phrase 'fair dealing' to qualify certain provisions, in practice, has allowed courts to introduce a variety of non-statutory and altogether new considerations into the calculus, even in circumstances where it might appear as though the legislature consciously sought to ensure against this eventuality.

The English Court of Appeal's observations in *Hubbard*, discussed earlier, have had an undue influence on the way in which Indian courts approach the question of fair dealing. Indeed, as commentators point out, English courts applying the statute's fair dealing provisions after *Hubbard* have failed to exercise the level of equitable discretion that Lord Denning tasked courts with in fair dealing cases.[2] Courts there have instead chosen to adhere to a strict construction of the provisions in question, largely unmoved by what their counterparts in the US have been doing under the rubric of fair use. This is, however, hardly the case in India.

Indian courts that have given sufficient thought to the issue have used Lord Denning's dicta in *Hubbard* to undertake a contextual scrutiny of the defendant's copying and measure its fairness against the consequences of allowing the copyright owner to prevent such copying. The case of *Civic Chandran* v. *Ammini Amma* (1996) is a good example. It involved a well-known dramatic work from which parts and scenes had been copied by the defendant, who went on to create and produce a 'counter drama' intended to 'to criticise the idea propagated by the drama and to expose to the public that the drama has failed to achieve the real object intended to be achieved by writing the same'(*Civic Chandran* v. *Ammini Amma* 1996: 684). When the plaintiff commenced a claim of copyright infringement, the defendant raised the defence of fair dealing under section 52(1)(a)(ii) of the act. Relying in large part on *Hubbard*, the Kerala High Court undertook a detailed comparison of the two works to find, eventually, that they had very different purposes and intentions and that allowing the copyright owner to obtain an injunction against the use would impact the defendant's free speech rights. Paying only limited attention to the text and legislative history, the court went on to allow the defendant's fair dealing defence. Central to the court's reasoning and holding was its conception of what fairness entailed, as allowed for by the 'fair dealing' requirement of the statute.

Other Indian courts have followed this trend and relied on the dictum in *Hubbard* to undertake a balancing approach to the question of fairness. A more recent and far-reaching trend in this same direction has, however, been some courts' willingness to treat fair dealing as entirely synonymous with fair use. The decision of the Delhi High Court in *The Chancellor*

[2] See Gowers (2006); Griffiths (2002).

Masters and Scholars of The University of Oxford v. *Narendra Publishing House and Ors.* (2008) captures this trend nicely. The plaintiff published several school textbooks, in particular one on mathematics written by a prominent Indian scholar that contained a series of questions for students. The defendants had published their own guidebook which sought to respond to the plaintiff's questions. It reproduced those questions and provided detailed answers/explanations to them—all without the plaintiff's permission. Consequently, the plaintiff brought an action for copyright infringement. In response however, the defendant claimed that his actions were protected under section 52 of the Copyright Act.

Here is where things became interesting. In dealing with the plaintiff's defence under section 52, the court began by characterizing it as the 'fair use' exemption. Lest it be thought that the court's characterization was merely semantic or nominal, the court then proceeded to detail the purposes of the fair use doctrine, quoting from the well-known article by Second Circuit Judge Pierre Leval (1990). It then proceeded to discuss the US case law on fair use, including *Folsom* v. *Marsh* (1841), *Universal City Studios* v. *Sony* (1984), and *Harper & Row Publishers* v. *Nation* (1985), to distil the core principles of fair use. Somewhat ironically, the court's analysis nowhere noted the reality that fair use in the US does today find mention in the copyright statute, which is in turn worded in vastly different terms from the Indian and UK statutes. After then citing English and Indian cases on the question of copyright exceptions, the court observed:

> The doctrine of fair use then, legitimises the reproduction of a copyrightable work. Coupled with a limited copyright term, it guarantees not only a public pool of ideas and information, but also a vibrant public domain in expression, from which an individual can draw as well as replenish. Fair use provisions, then must be interpreted so as to strike a balance between the exclusive rights granted to the copyright holder, and the often competing interest of enriching the public domain. Section 52 therefore cannot be interpreted to stifle creativity, and the same time must discourage blatant plagiarism. It, therefore, must receive a liberal construction in harmony with the objectives of copyright law. Section 52 of the Act only details the broad heads, use under which would not amount to infringement. Resort, must, therefore be made to the principles enunciated by the courts to identify fair use. (*OUP* v. *Narendera Publishing House* 2008: 396)

Not surprisingly, the 'principles' that it identified were the four factors from the US fair use cases, today codified in 17 U.S.C. §107. Applying them to the facts of the case, the High Court concluded that the defendant's use was indeed a fair use / fair dealing, thus exempt from copyright infringement by definition. Later opinions, especially of the Delhi High Court, have tended to follow the same approach.

This 'liberal' approach, in effect, blurs the divide between fair use and fair dealing elaborated earlier. It suggests that, from the simple reality that both approaches incorporate the notion of fairness into the analysis, Indian courts have come to treat them as interchangeable, disregarding the traditional dichotomy which English law had relied on and which continues to influence the approach of English courts to this day.

Some Indian courts have taken this liberal approach one step further and have read into the Copyright Act exceptions that find no statutory basis whatsoever. One of these exceptions finds no recognition even under the US fair use principles, where courts are more willing to depart from the language and structure of the statute. This is the defence of de minimis. Here, we return to the *Super Cassettes* case discussed earlier. The decision described in the previous section was handed down by a single judge of the Delhi High Court, and the matter was appealed to a full bench of that court, which overturned the decision. This was also seen in the case of *India TV Independent News Service Pvt. Ltd. & Ors. v. Yashraj Films Pvt. Ltd.* (2013). While it does not appear that the full bench overturned the single judge's approach to interpreting section 52, the panel nonetheless reversed the earlier decision by reading into the copyright statute an independent defence of de minimis copying.

The panel's decision begins with a reference to multiple US fair use decisions and the observation that 'even in India fair use is determined on the same four factors'. Nonetheless, the court observes that, quite independent of fair use, copyright law ought to accommodate minimal copying under the well-known legal maxim *de minimis non curat lex* ('the law does not concern itself with trifles'), especially in situations such as the one before the court, where the defendant had taken minimal amounts of protected expression from the plaintiff's protected works with no obvious economic harm. In its discussion of the issue, the court observed that the US courts had considered the question and rejected the adoption of a de minimis defence variously as part of the substantial similarity analysis, as part of fair use, and as a stand-alone exception. Nonetheless, the panel parsed the US fair use decisions to note that the de minimis idea was still at play, either as a part of the fourth fair use factor or as a separate factor of fair use.

To support its conclusion that the defence had not been affirmatively foreclosed in the US, the panel's opinion further cited an article authored by one of us, which analysed over 60 fair use opinions, to conclude that the fair use doctrine was sufficiently malleable to allow courts to decide cases on multiple, often non-statutory factors. The court used the conclusion of the article to infer that the fair use doctrine was indeed compatible with a de minimis defence. When also presented with a treatise on copyright law, again authored by one of us, which argues that there is no viable de minimis defence in the US, the court disagreed with the treatise on its interpretation of the US law and further suggested that the treatise writer's argument was based on his perception of an inconsistency in the case law, which certainly did not suggest the lack of a 'viable' de minimis defence. The opinion then proceeded to set forth why a de minimis defence is indeed necessary for copyright law and crafted a new de minimis defence against copyright infringement premised on a five-factor test, which it then applied to the facts of the case to conclude that the defendant's actions did not amount to copyright infringement.

What is telling about the court's opinion is its willingness to depart from the text and structure of the Indian Copyright Act, as well as the fortitude with which it undertook an analysis of the US copyright law, disagreeing with the commonly accepted views therein, and thereafter explicitly entered the realm of policymaking to create a new non-statutory exception.

To be sure, this decision is still an outlier in the Indian copyright landscape.[3] Yet, it vividly illustrates three important points for our analysis:

---

[3] Nevertheless, a continuation of this trend is to be seen in the even more recent decision of the Delhi High Court (Division Bench) in the Delhi University Photocopying case (*The Chancellor, Masters & Scholars of the University of Oxford* v. *Rameshwari Photocopy Services* 2017). The case involved an action for copyright infringement brought by several academic publishers against a photocopying

First, while the fair dealing approach suggests that courts are to confine themselves to interpreting and applying the text of the statute rather than making new law, the extent to which this mandate forms an actual constraint can vary from one copyright system to another. In England, fair dealing continues to constrain courts—but the same phenomenon does not seem to be the case in India. Second, whereas fair dealing is indeed structurally different from fair use, the reality is that fair dealing, too, gives courts equitable discretion in applying the provision, allowing courts to rely on variables and factors that, for the most part, track the US fair use jurisprudence. This explains how Indian courts have consciously conflated fair use and fair dealing in their analysis, relying on observations from *Hubbard*. Third, the extent to which the dichotomy continues to remain depends, it would appear, in large part on variables external to copyright law as such. Included under this rubric is the extent to which the norms of legislative supremacy and/or textual adherence influence judicial decision-making. In the UK, these norms are strongly entrenched and are routinely followed by courts. This is far from being the case in India, where the judiciary has taken an active role in policy-making and crafted Indian law in multiple areas. It was perhaps naive to expect that they would do otherwise in the realm of copyright law.

Patterns Despite Open-ended Factors: The Countervailing US Experience

The Congress has set forth four factors to consider in calibrating whether any particular usage qualifies as fair use, and the Supreme Court has ruled that all factors must be considered in any given case.

---

service at University of Delhi, which was entrusted with the preparation of course packs, put together by professors at the university. In finding the defendant's actions to be non-infringing, the court concluded that 'fair use' was an implicit part of the Indian copyright landscape such that unless the doctrine was shown to have been excluded by legislative intent, it ought to be 'read into the statute'. It then relied on the idea of fairness in use to conclude that the educational purpose being served by the course packs rendered the use a fair use. During its analysis, the court even acknowledged that its approach to fair use was distinctive (and more expansive) than what had been seen in the jurisprudence of American courts.

Therefore, as is frequently remarked, the determination of fair use in the US is, of necessity, ad hoc; an infinite variety of circumstances may affect the resolution of any concrete fact pattern.

Although true on paper, this summary fails to reflect all the experiences. The reality of applying fair use is that a variety of postures have arisen in which given exploitations have been vindicated—almost as if the legislature had specified in advance that this particular channel of exploitation should qualify as 'fair dealing'.

Let us begin with a narrow category—utilizations of copyrighted material that takes place in the context of a judicial proceeding, whether in the courtroom or in contemplation of a future court appearance. Consider *Bond* v. *Blum* (2003), an infringement case arising out of a child custody proceeding that unfolded in state court. Here, one William Slavin alleged that his ex-wife's present husband, William Bond, did not maintain a household suitable for the former's children with the ex-wife. To establish this, Slavin's attorneys introduced into evidence in the state court proceedings a manuscript authored by Bond describing how, at the age of 17, he planned and committed the murder of his father, fooled the police about his mental state, used the juvenile system to get off with merely a 'slap on the wrist', and then recovered the proceeds of his father's estate, all without any remorse. Bond explained in the manuscript, 'I wanted my father's money' (*Bond* v. *Blum* 2003: 393). That manuscript was autobiographical, inasmuch as Bond himself had been convicted in juvenile court of (and temporarily held in a mental facility for) the very crime described in *Self-Portrait of a Patricide: How I Got Away with Murder*. In response to Bond's complaint of copyright infringement, the defendants pointed out that they had no intent to exploit the book's manner of expression for any purpose whatsoever. The Fourth Circuit roundly vindicated fair use, labelling the infringement claim frivolous (and awarding the defence its attorney's fees as well).

Consider next *Healthcare Advocates, Inc.* v. *Harding, Earley, Follmer & Frailey* (2007), another infringement case arising out of underlying proceedings, this time for trademark infringement and trade secret misappropriation. After Healthcare Advocates initiated that underlying case, defence counsel Harding, Earley, Follmer & Frailey investigated the claims against its client by accessing archives stored on the Internet that chronicled the way plaintiff Healthcare Advocates had displayed its purported trademark on the date specified in the underlying complaint. In response, Healthcare Advocates filed a new case, alleging that the attorneys' investigation amounted to copyright infringement. This case likewise vindicated fair use by roundly rejecting the infringement claim: 'It would be an absurd result if an attorney defending a client against charges of trademark and copyright infringement was not allowed to view and copy publicly available material, especially material that his client was alleged to have infringed' (*Healthcare Advocates, Inc.* v. *Harding, Earley, Follmer & Frailey* 2007: 637).

One could multiply these examples. Parallel to the first instance of the murderer's autobiography, another case vindicated a police investigator's downloading of photographs from a victim's website to use in the course of a murder investigation (*Shell* v. *City of Radford, Va.* 2005) and a different case validated a TV show's display of the O.J. Simpson murder trial, even though it included a copyrighted photograph that had been admitted into evidence (*Kulik Photography* v. *Cochran* 1997). Parallel to the second instance of counsel defending a different case, another court vindicated the use of software produced by the adverse party during a mediation session as within the fair use defence (*Moran* v. *deSignet Int'l* 2008).

Nonetheless, there is one exception to the above rule: with respect to exhibits and demonstrative evidence commissioned for the express purpose of the subject litigation, it is unfair for the commissioning party to shirk its payment obligations and rely on the fair use doctrine to use the resulting products in court proceedings. After all, 'if works intended for use in litigation could be freely copied without payment to or permission from the copyright holder, there soon would cease to be any viable marketplace for such works' (*Images Audio Visual Prods., Inc.* v. *Perini Bldg. Co., Inc.* 2000: 1086).

Now, let us move to a category of wider application—comparative advertising. *Triangle Publications, Inc.* v. *Knight-Ridder Newspapers, Inc.* (1978) held reproduction of the cover from the plaintiff's copyrighted magazine fair use in a comparative advertisement for the defendant's magazine. *Sony Computer Entertainment Am., Inc.* v. *Bleem, LLC* (2000) arose in the context of defendant Bleem's 'software emulator', which allowed purchasers of game cartridges playable on Sony's PlayStation console to operate the games instead on a personal computer; it allowed

Bleem to feature on the packaging of its box a 'screenshot' from Sony's copyrighted game, showing the superior resolution achievable on a computer screen as compared to a TV set hooked up to the PlayStation. The Ninth Circuit emphasized that Bleem's emulator competed not with Sony's copyrightable software for its games but instead with Sony's hardware, namely the uncopyrightable PlayStation console. In fact, Bleem's product could even stimulate further sales of Sony's software (even if it dampened Sony's sale of its hardware console, a harm that stands outside copyright protection). In line with that sensibility, it should be added that the Federal Trade Commission has issued a ruling praising the social utility to consumers of truthful comparative advertising.

Moving beyond the realm of comparative to 'pure' advertising, *S&L Vitamins, Inc. v. Australian Gold, Inc.* (2007) extended the vector of fair use. The plaintiff in that case manufactured tanning lotions that were sold to salons under contractual provisions barring Internet resale; the defendant (who was not party to those contracts) obtained some of the plaintiff's product and offered it for sale over the Internet; its website included thumbnail pictures of the product, including its copyrighted label. The district court held that the only market usurped by the defendant's conduct was for the uncopyrighted tanning lotion, not for the copyrighted artwork on the label. It therefore validated as fair use the defendant's truthful advertising of products that it actually offered for sale.

As a final exemplar, let us encompass a very capacious purpose for which works protected by copyright can be used—to combat racism. Although the US Constitution, adopted in 1789, ratified the institution of chattel slavery that treated black slaves as less than human beings, the decisive event in the nation's history was the next century's Civil War, which abolished slavery through a series of constitutional amendments. No feature is more profound in the nation's consciousness than the resulting mandate of equal protection under law for all citizens. Remedial measures continue until the present day to overcome the stain from that founding sin of slavery. One such measure, driven by private effort and creativity rather than proactive state action, is the remodelling of literary works long considered classics in American society with a stark emphasis on the racist undertones in such earlier works. The case that follows portrays the tussle between such creative and purposeful transformation, and the copyright in the earlier work.

This case involved a challenge brought by the copyright owner of the largest seller of all time among literary works in the history of the US (short of the Bible, which stands outside copyright protection). In response to Margaret Mitchell's classic, *Gone with the Wind*, with its sprawling tale of life amongst the white gentry of the antebellum South (that is, those states of the US before the Civil War that relied on the institution of slavery), Alice Randall composed *The Wind Done Gone*, recounting the same plot developments revolving around the same characters, but told from the perspective of the enslaved underclass. For instance, where the former book recounts an amusing poker game in which a gentleman overplays his hand and ends up losing one of his minor slaves, in the latter's hands, the same events emerge as a heart-wrenching tale of a son wrested away from his mother's love into a hell of maltreatment. In countless other particulars, Randall 'exploded' the racist underpinnings of this classic work of American literature. The Eleventh Circuit validated the utilization as fair use (*Suntrust Bank* v. *Houghton Mifflin Co.* 2001). Its ruling stands in contrast to decisions of similar vintage from sister circuits. For instance, when a different author, going by the name 'Dr. Juice', employed the almost equally famous *Cat in the Hat* children's book to make a social commentary about the celebrated O.J. Simpson murder trial, the Ninth Circuit refused to accord the same 'parody' fair use defence that protected Alice Randall (*Dr. Seuss Enterprises* v. *Penguin Books USA, Inc.* 1997). When another author based a trivia-quiz book on the most famous TV show of the era, *Seinfeld*, the Second Circuit withheld the fair use imprimatur for the resulting *Seinfeld Aptitude Test* (*Castle Rock Entertainment, Inc.* v. *Carol Publishing Group* 1998). The latter two cases lacked the distinctive anti-racist cloak that protected and lent transformative character to the first instance.

But protection for copying undertaken to combat racism is not limited to bestsellers and other highly famous works. A small local police station put out a monthly newsletter with a tiny circulation limited largely to interested parties. One month, the bulletin included one officer's allegory contrasting people of various colours—the not-so-subtle subtext being that African Americans as a class constituted coddled criminals. A crusading newspaper reproduced the article in toto—of necessity thereby exposing it to a vastly larger readership. When the aggrieved officer tried to vindicate his copyright in the piece, the district court held that the public interest in viewing his exact expressions made it fair use for the defendant to reproduce the whole (*Belmore* v. *City Pages* 1995). It is hard to reconcile that ruling with the decision some years earlier from another court (*Salinger* v. *Random House, Inc.* 1987) that the public interest did not extend to seeing the exact language used in unpublished letters, even when only lightly quoted rather than reproduced in full—especially given that this case concerned celebrated author J.D. Salinger offering scathing commentary about Charley Chaplin, Oona O'Neil, and other figures of intense news value (as opposed to the entirely anonymous character of everything about the foregoing police newsletter). Yet that latter case denied the fair use defence that the former one vindicated. The difference, sub silentio, seems to be the overwhelming value of the desideratum of combating racism wherever it exists.

Another example unfolded when a radio talk-show host made veiled anti-Semitic comments. *The Boston Globe* ran a story about the contretemps. Not content merely to characterize what was said on the air, the newspaper went further and used taped excerpts to bolster its reportage. The defendant in that case likewise prevailed under the fair use doctrine (*National Association of Government Employees/International Bhd. of Police Officers* v. *BUCI TV, Inc.* 2000). Correlatively, another court validated the Council on American-Islamic Relations's copying of a conservative talk-show host's anti-Islamic statements by posting on its website a four-minute audio segment from the radio programme (*Savage* v. *Council on American-Islamic Relations, Inc.* 2008). By contrast, when the purpose of copying was not to combat racial animus but merely to expose the liberal bias of the mainstream press by reproducing articles that appeared there, the fair use defence failed (*Los Angeles Times, Inc.* v. *Free Republic* 2000). Also worth mentioning here is the ruling in favour of the TV show *South Park*, when it substituted new lyrics for the famous Disney song 'When You Wish upon a Star' to become 'I Need a Jew'. The court derived support for its fair use ruling from the usage's dig at Walt Disney personally, given a 'widespread belief' that he was anti-Semitic (*Bourne Co.* v. *Twentieth Century Fox Film Corp.* 2009).

Of course, the three categories just confronted are not themselves exhaustive. Commentators have posited a variety of circumstances that effectively prevail, on a consistent basis, as fair use. Included here are perversion of copyright law as a pretext to suppress free speech on issues of public importance, quotation by a documentarian or by a historical work of snippets from prior works in order to advance a thematic point or to set a historical context, and the development of technologies designed to facilitate personal fair uses, as well as search-engine copying for the purpose of indexing or otherwise making information about protected works more publicly accessible.

Based on the foregoing, it would seem that, as a matter of legal realism, the US copyright law embodies the following non-exclusive categories of ersatz 'fair dealing':

1. copying for purposes of judicial proceedings, except with respect to material commissioned for the purpose of that proceeding;
2. copying for purposes of comparative advertising; and
3. copying of expressive materials on account of racist components reflected therein.

***

To summarize, in theory and when viewed in the abstract, 'fair use' and 'fair dealing' are routinely described as two competing approaches to the delineation of limitations and exceptions in copyright law. Legal systems are thus seen as needing to make a choice between them and, in the process, having to trade off competing structural considerations. While this may be true as a matter of principle, a comparison of the Indian and US approaches reveals that the story is far more nuanced than what the simple dichotomy suggests. In practice, and as the system develops, the flexibility and open-endedness of fair use can indeed produce recurring patterns, while, conversely, the rigidity of fair dealing might indeed also produce new lawmaking from courts. The set of background principles motivating legal analysis and the institutional dynamics between courts and other legal institutions in the system seem to play an equal (if not more important) role in determining the direction that each approach takes. Consequently, one needs perhaps to be a bit more circumspect before predicting where either approach will lead a copyright system and its approach to innovation.

Cases Cited

Authors Guild, Inc. v. HathiTrust, (2014) 755 F.3d 87 (2d Cir. 2014).

A.V. ex rel. Vanderhye v. iParadigms, LLC, (2009) 562 F.3d 630 (4th Cir. 2009).

Belmore v. City Pages, (1995) 880 F. Supp. 673 (D. Minn. 1995).

Blanch v. Koons, (2006) 467 F.3d 244 (2d Cir. 2006).

Bond v. Blum, (2003) 317 F.3d 385 (4th Cir. 2003).

Bourne Co. v. Twentieth Century Fox Film Corp, (2009) 602 F. Supp. 2d 499 (SDNY 2009).

Cambridge Univ. Press v. Becker, (2012) 863 F. Supp. 2d 1190 (ND Ga. 2012).

Cambridge Univ. Press v. Patton, (2014) 769 F.3d 1232 (11th Cir. 2014).

Campbell v. Acuff-Rose Music, Inc., (1994) 510 US 569.

Carla Calkins v. Playboy Enterprises Intern., Inc., (2008) 561 F.Supp.2d 1136 (E.D. Cal. 2008).

Castle Rock Entertainment, Inc. v. Carol Publishing Group, (1998) 150 F.3d 132 (2d Cir. 1998).

Civic Chandran v. Ammini Amma, (1996) 16 PTC 670 (Ker).

Columbia Broadcasting System, Inc. v. Loew's Inc, (1958) 356 U.S. 43.

Dr. Seuss Enterprises v. Penguin Books USA, Inc., (1997) 109 F.3d 1394 (9th Cir. 1997).

Folsom v. Marsh, (1841) F. Cas. 342 (CCD Mass. 1841).

Harper & Row Publishers v. Nation, (1985), 471 US 539.

Healthcare Advocates, Inc. v. Harding, Earley, Follmer & Frailey, (2007), 497 F. Supp. 2d 627, 639 (ED Pa. 2007).

Hubbard v. Vosper, (1972) 2 QB 84 (Eng.).

Images Audio Visual Prods., Inc. v. Perini Bldg. Co., Inc., (2000) 91 F. Supp. 2d 1075 (E.D. Mich. 2000).

India TV Independent News Service Pvt. Ltd. & Ors. v. Yashraj Films Pvt. Ltd., (2013) PTC 586 (Del).

Kulik Photography v. Cochran, (1997) 975 F. Supp. 812 (ED Va. 1997).

Los Angeles Times, Inc. v. Free Republic, (2000) 54 USPQ.2d (BNA) 1453 (CD Cal. 2000).

Moran v. deSignet Int'l, (2008) 557 F. Supp. 2d 378 (WD NY 2008).

National Association of Government Employees/International Bhd. of Police Officers v. BUCI TV, Inc., (2000) 118 F. Supp. 2d 126 (D Mass. 2000).

Pro Sieben Media AG v. Carlton UK Television Ltd, [1997] EMLR 509.

S&L Vitamins, Inc. v. Australian Gold, Inc., (2007) 521 F. Supp. 2d 188 (EDNY 2007).

Salinger v. Random House, Inc., (1987) 811 F.2d 90 (2d Cir. 1987).

Savage v. Council on American-Islamic Relations, Inc., (2008). 87 USP.Q.2d (BNA) 1730 (ND Cal. 2008).

Shell v. *City of Radford, Va.*, (2005). 351 F. Supp. 2d 510 (WD Va. 2005).

Sony Computer Entertainment Am., Inc. v. *Bleem, LLC*, (2000) 214 F.3d 1022 (9th Cir. 2000).

Stewart v. *Abend*, (1990) 495 U.S. 207.

Suntrust Bank v. *Houghton Mifflin Co.*, (2001) 268 F.3d 1257 (11th Cir. 2001).

Super Cassettes Indus. Ltd. v. *Chintamani Rao*, (2012) 49 PTC 1 (Del).

The Chancellor Masters and Scholars of The University of Oxford v. *Narendra Publishing House and Ors.*, (2008) 38 PTC 385 (Del).

The Chancellor, Masters & Scholars of the University of Oxford v. *Rameshwari Photocopy Services*, 2017 (69) PTC 123.

Triangle Publications, Inc. v. *Knight-Ridder Newspapers, Inc.*, (1978) 445 F.Supp. 875 (S.D. Fla. 1978).

Universal City Studios v. *Sony*, (1984) 464 US 417.

Walt Disney Productions v. *Air Pirates*, (1978) 581 F.2d 751 (9th Cir. 1978).

Williams & Wilkins Co. v. *United States*, (1975) 420 U.S. 376.

References

Gowers, Andrew. 2006. *Gowers Review of Intellectual Property*. London: HMSO Licensing Division. Available at http://webarchive.nationalarchives.gov.uk/+/http:/www.hm-treasury.gov.uk/d/pbr06_gowers_report_755.pdf, last accessed on 29 March 2017.

Griffiths, J. 2002. 'Copyright Law after Ashdown: Time to Deal Fairly with the Public', *Intellectual Property Quarterly*, 3: 240.

House of Representatives. 1976. *Copyright Law Revision*, Report No. 94-1476. Available at https://www.copyright.gov/history/law/clrev_94-1476.pdf, last accessed on 29 March 2017.

Latman, Alan. 1958. *Fair Use of Copyrighted Works*, Study No. 14, Studies Prepared for the Subcommittee on Patents, Trademarks, and Copyrights of the Committee on the Judiciary United States Senate (86th Congress, 2nd Session). Available at https://www.copyright.gov/history/studies/study14.pdf, last accessed on 29 March 2017.

Leval, Pierre N. 1990. 'Toward a Fair Use Standard', *Harvard Law Review*, 103(5): 1105–36.

Nimmer, David. 2003. '"Fairest Of Them All" and Other Fairy Tales of Fair Use', *Law and Contemporary Problems*, 66(1): 263–87.

Nimmer, Melville B. and David Nimmer. 2012. *Nimmer on Copyright*. New York: Matthew Bender.

7

T. SUNDARARAMAN
RAJANI VED

Innovations in the Organization of Public Health Services for Rural and Remote Parts of India

In 2011, the Government of India created a Sector Innovation Council for the health sector to study innovation in this sector and recommend future policy. The council's mandate was not only to evolve an innovation-friendly ecosystem but also align innovations in this sector with the public health priorities and national goals. Over a year, the council developed a comprehensive report, which has since been published by the National Health Systems Resource Centre (NHSRC) (Sundararaman 2012). The council examined innovation in four different contexts: pharmaceuticals; medical devices; health informatics; and public health service delivery. This chapter draws extensively upon the council's study of innovations in the delivery of public health services.

Innovations in the Discourse on Health Sector Reform

Innovation is only one of several contributors to a better-performing public health service system. The most important determinant is clearly political commitment to increasing public investment in health and public health expenditures. The availability of human resources and the quality of governance are also important determinants of performance. Indeed, so overwhelming are the roles of these other contributors that the role of innovation has seldom been recognized and often ignored.

A major reason for the under-appreciation of innovation is that many equate reform with the state ceding space to private actors—a product of health sector reform discourse in the 1990s. This often meant the state withdrawing to handling only a few high-priority items and letting the market handle the rest. Problems with service delivery were to be addressed through better enforcement of the rules. What little mention there was of innovation was restricted mainly to forms of public–private partnership (PPP).

While the initial years witnessed some improvement in reproductive and child health indicators, there was no further increase in the quantity, quality, or efficiency of public services. It was only in 2005, with the launch of the National Rural Health Mission (NRHM), a renewed emphasis on public investment in healthcare, and a greater role for the federal government in strengthening health systems, that innovation as a means of improved public health service delivery entered the discourse. Though PPPs continued as significant actors in the innovation story, the trend to treat the terms 'PPP' and 'innovation' as almost synonymous declined.

Defining and Recording Innovation in the Context of Health Systems

The Sector Innovation Council considers a change to qualify as 'innovation' only if it '(a) meets a need or solves a problem; (b) is creative and involves a new approach or a new application of an existing approach; and (c) brings significant benefit to one or more groups' (Sundararaman 2012: 105). Innovations related to service delivery can be a comprehensive business model or deal with select elements of the implementation chain. The revival of interest in health-sector innovations over the past decade and a half has led to the emergence of quite a few databases that compile and disseminate information on this topic. The driving rationale was that once these best practices are collated in an easily accessible format, their wide dissemination would assist other governing authorities in picking them up and improving their own performance. Most of the recorded innovations are community processes or PPPs, in line with the thinking of that period.

Some of the important databases are:

1. The European Union's (EU) Sector Investment Programme has created a database which is managed by the Central Bureau of Health Intelligence. It has successfully compiled about 208 entries, of which some have been scaled up effectively.
2. The United States Agency for International Development (USAID) database of 36 entries focuses on community-level interventions and those that involve PPPs. The USAID-funded Vistaar database also conducted an evidence review of innovations in maternal health, infant and child health, and nutrition.
3. A directory compiled by the Indian Ministry of Health and Family Welfare (MoHFW) and supported by the Department for International Development (DFID) of United Kingdom (UK), includes seven categories and 229 entries. Most innovations in this database, which considerably overlaps with earlier databases, relate to reproductive and child health programmes and the NRHM. This was completed in 2008.
4. Another database is the Health Market Innovation Directory, compiled in 2011 by the Centre for Health Market Innovation—an organization which is part of the Washington-based Results for Development Institute. This has profiled 1,015 innovations across 107 countries. In this list, 215 are from India.

Most of the innovations from India are led by non-governmental organizations (NGOs) and financed by donors. Very few of the innovations in this list have gone to scale. The few that are on scale are government-led and did not begin as small-scale innovations.

In parallel to these database efforts, the NHSRC, acting as a technical arm to the NRHM, has organized Best Practices and Innovations conferences. Four such conferences have been held thus far. These conferences have attempted to identify the most intractable problems of public health services delivery common to the various states of India, and then invite presentations from states that were relatively more

successful in addressing these problems. Often, states had not categorized their own efforts as an 'innovation' or were not even aware that
they were doing relatively better with respect to these problems. The
NHSRC's state observers have thus played an important role in bridging the information gap and writing up case studies that facilitate best
practices adoption.

Therefore, the four NHSRC conferences have been certainly productive in documenting innovations that already operated on scale and in
bringing about higher success rates of adoption through cross-learning
and replication. Their lasting contribution, however, is in the way they
have transformed thinking about the problems that ail public health services delivery in India. They have substituted the view that these problems are intrinsic and unavoidable collaterals of any initiative of a public
character with the belief that innovation can infuse administrative and
technical competence into their conceptualization and execution. The
Sector Innovation Council has provided an analytic framework to categorize and understand innovations in services delivery, and has also
recommended optimal ecosystem requirements for an innovation-
friendly health system.

Innovation Pathways

There are, broadly speaking, three pathways for structuring innovation
in public health services delivery:

1. *The best practices pathway*: Here, innovations and solutions are developed at a local level as a planned effort by a knowledge agency or
   local administrators. The innovation gets noticed and is considered
   for scaling up; not all successful innovations readily scale up. Some
   are better seen as local adaptations, relevant only to that context.
   Others are pilots, often driven by aid agencies based on their international experience, but may fail to convince the targeted public. Yet
   other successful small-scale innovations subsequently get subsumed
   under new policy initiatives from above.
2. *The business model pathway*: This pathway usually involves corporate or
   private sector initiatives calculated to address an existing need or problem while generating profit by creating a market for the innovation.

3. *The policy priority pathway*: This pathway involves the government developing and scaling up new schemes in response to policy priorities.

We present here a few case studies that reflect each of these approaches, all of which are taken from the Sector Innovation Council report (Sundararaman 2012).

Best Practices Pathway Case Studies

The Gadchiroli Model of Home-based Newborn Care: Scaling Up to a National Level

This innovation is a classic example of how a small-scale pilot, to address the specific issue of high neonatal mortality, implemented in 30 villages of a resource-depleted district in Maharashtra, was scaled up to the entire country. The originator of the innovation in this case was the Society for Education, Action, and Research in Community Health (SEARCH), an NGO led by a couple, both public health professionals with specialist clinical skills who were inspired and guided by Gandhian ideals. The essence of the innovation was the provision of newborn care within the home through a trained community health worker (CHW), resulting eventually in a 60 per cent reduction in neonatal mortality. The genesis of the innovation lay in the following facts: the most vulnerable period for sickness and death is the first week of life to the first month of life; recognizing illness requires a certain skill set; and facility-based care is beyond the geographic or economic means of poor rural families.

India's massive CHW programme, the ASHA programme, incorporated this approach and scaled it up nationwide. ASHA is an acronym for accredited social health activist and was also chosen since *asha* in Hindi means 'hope'. The ASHA was usually a literate woman, resident of the locality, selected by the community, trained and deployed as a community volunteer, who drew a performance-based monetary incentive for some tasks but no regular salary. As the programme evolved, the ASHAs were never formally accredited, and their activist character was overshadowed by their functions in service delivery and facilitating access to services. The name ASHA stuck however, and now signifies a woman CHW generally. Three key elements that appear to have led to the success of the programme were the recognition that, despite increases in institutional delivery: (a) a substantial proportion of newborns die at home; (b) the availability of 800,000 ASHAs who provided the delivery system for the innovation; and c) the learning and confidence obtained from the experience of the Mitanin programme, detailed later in this chapter. The innovation was scaled up first by other NGOs, and then through state support. The efforts led to effective reductions in neonatal mortality, and the results were published internationally and disseminated and championed at various national fora. Home-based newborn care was listed as a major component of the Eleventh Five Year Plan and a key strategy for reduction in infant mortality rate. From the first pilot to its successful nationwide scaling up, the process has taken nearly 20 years.

The Purulia Special Newborn Care Unit (SNCU)

In 2003, with the support of UNICEF, the neonatology department of the B.C. Roy Medical College in Kolkata took up the development of an intensive newborn care unit in the Purulia district hospital. Faced with a human resource crunch, they had to upgrade the skills of existing medical officers and develop a cadre of para-nurses to supplement the scarce available nursing strength. Using appropriate intensive care practices and protocols, they demonstrated that neonatal mortality could be substantially reduced through an appropriately designed SNCU.

Scaling up this model took considerable time. This model was already well established in 2005 when the external-aid-supported Reproductive and Child Health programme Phase 2 (RCH-II) was being designed. The Purulia model was not included within the ambit of RCH-II because the government expected that it could successfully recruit private providers to perform this role without having to develop its own outfit. It was only in 2009, when the government recognized the lack of a referral site for sick newborns, that it became an integral part of the programme. Since then, it has rapidly increased to more than 200 units. This case study illustrates an innovation in the organization of facility-based public health services and how scaling up requires both a viable model and a favourable policy environment.

Referral Transport for Malaria Slides

A key constraint of India's malaria control programme is to ensure timely examination of blood slides (from the village) to better guide treatment and public health action. The average 7–14 days between the time the peripheral worker takes a blood smear in the field and obtains a report is often too late for either improved clinical care or public health action.

The Jan Swasthya Sahyog (JSS), an NGO working in the district of Bilaspur in Chhattisgarh, has a field area spread over an entire block, with its laboratory and headquarters located some 10 km from the town of Bilaspur. The JSS arranged with village volunteers to send the blood smears to the nearest local bus stop. The bus staff were instructed to deposit the smears with the laboratory. The report, once ready, would follow the reverse path and get to the village volunteer by the same evening.

This innovation never really scaled up, mainly because of the introduction of rapid diagnostic kits. Moreover, the government's ability to work on a participatory model at all levels is limited. This case study is important because it is a local solution that was recognized as important and selected for scaling up but programme dynamics did not favour it. Yet, it remains relevant and illustrates a community-based innovation for improving service delivery.

Referral Transport in Nabrangpur, Odisha

A key barrier to institutional delivery is the availability of affordable, readily accessible transport that can take expectant mothers from home to hospital within a maximum of 30 minutes. Though most states have introduced a centralized 'Dial 108' ambulance service approach, much on the lines of the 911 service in the United States (US), this system is much less effective in reaching rural and remote areas. However, an evaluation of the Janani Suraksha Yojana, led by the NHSRC (Ved et al. 2011), drew attention to the fact that in the remote, conflict-ridden district of Nabrangpur in Odisha state, nearly 68 per cent of all pregnant women who accessed institutional delivery had used a locally designed and operated, and publicly financed patient transport service. This was a modification of the Janani Express service, a model tried with limited success in a number of states. This level of utilization is more than twice the highest utilization rates in other districts studied.

The service relied on a systematic mapping of villages with specific facilities, positioning one vehicle each that could respond to calls from most villages within half an hour. It also relied on contractual arrangements for the provision of vehicles and drivers, with a payment model built on a fixed cost for retention and a variable cost for the actual usage. Users of the service, however, paid nothing. The sharing of mobile numbers between the facility coordinator, the community health volunteer, and the service users facilitated immediate response to the call. The local drivers knew the terrain and could negotiate their way across the forested, conflict-ridden area more efficiently. Utilization was also high because there was no alternative transport available; women, motivated by the community health volunteers, did desire to move to the clinic for their deliveries. The Janani Express scheme is an example of an innovation by programme managers who used locally available solutions to guarantee dependable referral transport. It is also an example of how a best practice can be used to understand and overcome seemingly intractable barriers to service delivery.

The Boat Clinics of Assam

The Brahmaputra River, which cuts across Assam, is massive, with large, well-populated islands that are sequestered from most public health services delivery systems. There is no private sector penetration in these islands and the population is completely dependent on public health services for even basic curative care needs. However, public health providers, especially doctors, are reluctant to reside in these islands. The solution for reaching services here was through mobile medical units organized in the form of suitably equipped and staffed boat clinics.

This innovation is an adaptation of the mobile medical unit to a specific context; and it offers useful learning, with considerable potential for replication in similar contexts.

Voucher Scheme to Increase Institutional Delivery: Haridwar, Agra, and Kanpur

This innovation traces its genesis to an internationally promoted model of PPP built around vouchers. These vouchers, issued to women in the below the poverty line (BPL) category, entitle them to a set of maternal health or newborn care services of high quality. Private providers honour these vouchers without any payment from the beneficiaries and are later reimbursed by the state government. This was expected to increase the involvement of private sector providers in publicly financed service delivery. The NGOs were appointed to coordinate between the public and private sectors, to supervise and monitor the quality of care, and to train and supervise the ASHAs. The vouchers (serially numbered, with holographic stickers to prevent counterfeiting) were provided to pregnant women through a chain involving NGOs and ASHAs. District chief medical officers provided supervision, and the project funds facilitated reimbursement.

Pilot programmes were first launched in the cities of Haridwar, Agra, and Kanpur. A review of the scheme one year after implementation demonstrated several problems related to the high proportion of caesarean deliveries, long waiting periods, difficulties in segregating the BPL patients, and variable quality of services in the private hospitals. Part of the poor quality stemmed from limited monitoring and supervision of the private hospitals, lack of mechanisms to redress grievances, and lack of newborn care services. Due to such 'operational problems', the programme's success was never established and it never got the support needed to scale it up. Whereas most of the other innovations were 'incremental innovations', addressing very specific problems, the effort in this case was a 'disruptive innovation', to develop an alternate approach not only to financing but also to the organization of health service itself, which perhaps contributed to the inability of this voucher scheme to take off in a larger way.

Business Model Pathway Case Studies

'Dial 108' Services

Under the auspices of Satyam Computers, a corporate firm, the Emergency Management and Research Institute (EMRI) was set up as a company to deliver emergency response services modelled on the 911 service in the US. A call centre, reachable by dialling 108, would respond to all emergencies, fire, crime, or medical, by diverting calls to the fire department, police, or ambulance services, respectively. Within 20 minutes of a medical emergency, the ambulance would reach the patient, a trained emergency paramedic would administer first aid, and the ambulance would transfer the patient to a private or public hospital in the vicinity. Satyam started it with an intention to explore building a model based on cost recovery, but shifted to a publicly financed venture once it was clear that a paid model was unlikely to work. The state government, supported financially by the NRHM, reimbursed the expenses incurred, but with no cap on costs. The planning, technical know-how, and software used at the call centre were all provided by Satyam. During the course of the programme, corporate ownership shifted to the GVK Group. The government played an active role in finding another corporation to take over this role on a not-for-profit basis.

Evaluation showed high public appreciation for the availability of a service which had hitherto been unavailable. It also showed that they only used the service in about 30 per cent of emergencies—the more distant and vulnerable the population, the less likely they were to access it. It was most effective in trauma care. It was also effective for transport of pregnant women, though cheaper alternatives may have been more efficient. More vehicles were needed to provide truly timely coverage; the costs, already rising, would have become prohibitive. The model shifted to hiring providers through a public tendering process, which allowed for a few more players and better ways of calculating costs and improving service delivery.

This adapted model has been hugely successful. An estimated 23 states have now established a publicly financed emergency response system, and over 11,000 ambulances provide services, responding to more than 35,000 medical emergencies per day. In terms of value for money and deployment of state-of-the-art technology, it has few equivalents anywhere in the world. This is easily India's largest and most successful PPP in the health sector, both in terms of public expenditure and service delivery.

The Fixed-day 104 Services

This innovation was also piloted by Satyam Computers in Andhra Pradesh; after the fall of Satyam as a corporate entity, the management of this project became vested in an autonomous company. Here, the model consisted of four interrelated components: (a) a 104 medical helpline for telephonic medical consultation; (b) a mobile medical unit to provide medical consultation through paramedics; (c) a training programme for informal, unqualified providers working in rural areas (also known as registered medical practitioners, or RMPs, though they are not registered); and (d) a telemedicine link. This project explored an alternative approach to the delivery of primary care services. Its attempt was primarily to entrust paramedics and RMPs with medical care and to provide, where required, medical first-referral support through vehicular fixed-day services and secondary care support through telemedicine links. The mobile medical unit was geared to detect and offer follow-up treatment for common non-communicable diseases (NCDs) and equipped with adequate drugs to do so.

The NCD component worked and made drugs for the poor accessible, but other than this, neither were the actual gains and the cost–benefit ratios clear, nor were health outcomes measurable. Most importantly, the financing of this model, as with EMRI of the previous case study, was on the 'whatever it takes' principle. The difference was that, whereas EMRI offered a supplementary service that strengthened utilization and effectiveness of public health services, this project was visualized as a substitute for sub-centres and primary health centres. In this model, a user fee was charged by the RMP or paramedic and by the specialists or hospitals who treated the referred patients. The call centre and mobile unit were free for the user but paid for by the government.

By early 2011, the programme had lost governmental support. Its workers went on strike, demanding absorption as regular government employees, and the programme eventually closed down. Faced with increasing costs and no clear outcomes, the government took over the scheme. It now functions in the form of mobile medical units elsewhere—as an outreach service linked to the primary healthcare network.

Janani and the Merrygold Variant

Janani is an NGO in Bihar working to reduce the high unmet need for quality family planning services through a combination of its own network of clinics, franchisees and clinics in the private sector, and the social marketing of branded contraceptives. Janani provides family planning guidance and abortion services, charging a user fee for the latter.

The most innovative part of this programme is the social franchise component, wherein Janani recruits private clinics to provide a standard package of core reproductive and child health services. Costs and quality are standardized, with a clear fee for services. A robust monitoring system enforces this. In return for the franchise fee, the clinic gets management and training support and a volume of patients. Linkages with government programmes where government reimburses the cost of care, for example, in sterilization surgeries, help bring in substantial volumes and income. The revenue model of this programme is thus built on safe abortion and sterilization services, offered for a fee to patients who access its services directly and free of cost for patients referred by the government. The business model is such that, even without any link to the government, this network of hospitals would provide affordable and ethical care for most of the people. In such a setting, the government could supplement the network's effort by purchasing some select and well-defined services or by paying for treatment of the poorest. It is worth noting that a substantial proportion of surgeries for controlling births in Bihar were being offered by this network.

There are many problems in this programme. Franchisees often build their clientele and business network, then leave and create their own private hospital networks. The proportion of clinics in the network which are directly invested in, owned, and managed by Janani has increased, but there are limits to such growth in a scenario where the regulation of the rates and quality of care is very weak.

Merrygold was an effort to replicate a similar social franchise concept in Uttar Pradesh. It was funded by USAID and implemented through the Hindustan Latex Family Planning Promotion Trust (HLFPPT), a public sector undertaking. Again, the search was for a form of privately or autonomously owned service delivery from which government would purchase services. Because of its USAID funding support, it excluded safe abortion services from its offerings. The response has not been as positive as with Janani, though a few peri-urban facilities are doing good business. However, since alternative private sector providers are available in these areas at comparable rates, the value addition is not as clear.

Aravind Eye Hospitals

The Aravind hospitals are a private sector hospital chain for eye care. They sell a well-defined service—cataract surgery—to the government. However, as with Janani, this is a very small part of its business. What is special about this chain is that it provides highly subsidized—almost free—care for poor patients. For all patients, the cost is very modest, but the paid part of the business model subsidizes free care for the poor. The poor get the same quality of medical care, but hospitality for paid in-patients in the category is better, with the option to follow up with a doctor of one's choice.

This business model requires a very high level of efficiency, and the key factor is a carefully maximized turnout of patients. Such efficiency is gained by training paramedics in-house for all non-critical procedures, leaving only surgery and final diagnosis to specialists. This can result in an approximately eightfold increase in the number of surgeries a doctor performs. By carefully leveraging government schemes for cataract surgery and developing its own outreach services, the central hospital performs over 300,000 eye surgeries every year, nearly half of which are free. This model is essentially independent of government support, though governmental purchase of cataract services helps.

The Chiranjeevi Scheme

Chiranjeevi Yojana was started in Gujarat to contract maternity services to the private sector. Here, the government reimbursed private providers for providing delivery and emergency obstetric care services. The contracting package was innovatively designed and implemented, bundling complicated cases and normal delivery option into a single contract.

There were initial concerns with monitoring the provision of cashless services and quality of care. Moreover, the contracted services were within a very narrow range. So, when private practice in rural settings became more viable with this scheme, a number of doctors left public service for the private sector, which weakened the public sector's ability to cope with the rest of the healthcare needs it faced. In turn, the private sector did not see it very useful to remain within the contracting arrangement; many left the contract and began handling most deliveries without any form of insurance or financial protection.

Several states have replicated this innovation: examples include the Saubhagyawati Scheme (Uttar Pradesh), Janani Suvidha Yojana (Haryana), Janani Sahyogi Yojana (Madhya Pradesh), Ayushmati Scheme (West Bengal), Chiranjeevi Yojana (Assam), and Mamta Friendly Hospital Scheme (Delhi), but in these states the scale is nowhere near Gujarat. It is at best a minimal, though useful, supplement.

Rashtriya Swasthya Bima Yojana

The Rashtriya Swasthya Bima Yojana (National Health Insurance Programme) is one of the most important innovations of the 2000s. This is a social insurance model. The government recruits an insurance service provider through a tendering process to deliver a standardized insurance package. The sum assured is INR 35,000 per year for a family of five; the premium is decided by the bid, but is in the range of INR 700; and the co-payment by the insured for enrolment is only about INR 30. Coverage is only for hospitalization. Both public providers and private providers can be empanelled.

The reach of this model has increased rapidly to over 120 million individuals, making it the largest and most successful programme to date for purchasing care from private providers. However, objective evidence that the programme is providing social protection against catastrophic illness is limited, and there is a trend of both unnecessary care and unnecessary costs. In most states, there is a low claims ratio, but where there is public awareness and providers are available (for example, in Kerala), everyone tends to claim the sum assured and the insurance company takes a beating. In some states, public sector institutions account for most claims, because this scheme provides a more flexible and responsive financing mechanism. A major concern has been that because much of the care provided is supply driven, the poor may be unable, in times of genuine medical emergency, to secure their entitlements under this scheme. There is, therefore, a call to reform this scheme and improve its implementation. As part of this effort, it has been transferred from the Ministry of Labour and Employment to MoHFW on 1 April 2015.

The Society for Elimination of Rural Poverty (Velugu) Model

The Society for Elimination of Rural Poverty model is unique in that it is based on 'social enterprise' principles and addresses, almost entirely, both preventive and promotive healthcare efforts. No other business model has ever been tried in both preventive and promotive health. This project extends to about 600 villages in Andhra Pradesh.

The programme advances a holistic approach to tackle concerns of nutrition, women's empowerment, maternity entitlements and incomes, and rural employment. The village community and self-help group jointly manage a kitchen with trained cooks, who supply two meals to pregnant women and young children at a subsidized charge. Pregnant women are offered income-generating work that shifts them from farm labour to more sedentary activity, while also paying for the increased nutrition requirement. The kitchen is open to other women and even men, on payment at reasonable rates. Eventually the entire cost of the diet is recovered over two years, which implies that the system could recover most of its running costs. Only the initial capital and capacity-building costs and some subsequent management costs are external.

The innovators behind this programme were senior civil servants with a track record of working with pro-poor developmental schemes. The financing was through external assistance, mainly from the World

Bank. This was perhaps due to the flexibility that such external assistance offered and the World Bank's own preference for social enterprise models. Scaling up has, however, remained a challenge since the perception is that this approach requires much higher levels of institutional capacity and social capital than most village communities can offer.

Policy Priority Pathway Case Studies

Tamil Nadu Medical Services Corporation (TNMSC)

One of the first and most successful of all health system innovations in the Indian context, the TNMSC was pioneered by the Government of Tamil Nadu in 1996. The corporation put in place an institutional structure and organization for procurement and supply chain management of essential drugs and supplies to all public health facilities, getting the best rates and a high degree of quality assurance in the process. Much of its success can be attributed to getting the supply chain right—there were no stock-outs in any facility and the supply was responsive to the demand pattern. All this was achieved without any sophisticated digitized inventory system.

One measure of the programme's success is the much higher level of financial protection it provides for its users, with the state showing only about one-tenth of the national average for out-of-pocket expenditure during hospitalization in a public hospital. A recent study confirms that the average (taken as median) medical expenditure for out-patient, in-patient, or delivery care in the district healthcare system is zero. A number of states, notably Rajasthan, are now replicating this success.

Retention Schemes for Skilled Health Professionals

The Eleventh Five Year Plan (2007–12) despaired of difficulties in getting doctors to work in rural areas. By the time of the Twelfth Plan (2012–17), this was seen as a problem that could be solved with administrative competence. This was due to the large number of innovations that showed this problem to be tractable. The measures taken since 2005 include:

- better financial incentives for working in difficult areas, which has worked well in Himachal, Chhattisgarh, Odisha, and a number of other states;
- non-financial incentives, especially preferential admission to the limited opportunities for postgraduate education;
- innovations in the recruitment process, such as walk-in interviews, tried in states like Haryana, Jharkhand, and Maharashtra;
- giving preference to candidates from under-serviced areas;[1]
- special courses leading to a new cadre dedicated to rural service, like the three-year rural medical assistant course in Chhattisgarh and Assam;
- multi-skilling medical officers through short-term courses in specialist skills for emergency obstetrics and anaesthesia, to close the gaps in emergency obstetric care in rural and remote areas;[2]
- distance-education-based approaches to building up necessary skills in situ for epidemiologists, family medicine, district health management, and more; and
- creating a separate cadre with an extra-benefits package to promote rural service, which has worked very well in Chhattisgarh.

It is important to learn from these diverse experiences. Platforms for sharing these innovations across states have encouraged states to learn from others and initiate their own experiments and models.

Mitanin and ASHA

The Mitanin programme is an innovation by the Chhattisgarh state government to improve health outcomes through an appropriate process

---

[1] This works even better where communities are actively involved in selecting and sponsoring local men or women for professional education and deployment, with the understanding that they will come back to serve the local community: for example, a second auxiliary nurse-midwife was selected, trained, and deployed for each of the health sub-centres of West Bengal.

[2] This has been tried in all states. Its effectiveness varies depending on conditions.

of selecting, training, and deploying community health volunteers. *Mitanin*, in the local language, Chhattisgarhi, denotes a friend, usually of the same sex and age group, with whom another girl child, usually of school age, is bonded through a traditional ceremony. The state chose to name their women community health volunteers 'health mitanins' since the word invokes the voluntary service in times of distress that is expected of one's mitanin. In this context, the health volunteer is the village's mitanin.

The programme was scaled up in phases from 2003 to 2005 and now covers every hamlet in Chhattisgarh, with about 54,000 mitanins in place. The premise of the programme is not new. Situating one mitanin, trained in a health rights framework, to provide community-level care and facilitate access to services, in every village and town, is an approach that has worked in many small-scale NGO programmes. But when scaled up and led by government, it had failed in the past. The Mitanin programme not only managed to scale up but was also the inspiration for a nationwide scaling up to the ASHA programme, which is strong and growing with over 850,000 mitanins. A number of evaluations over the last 10 years have shown considerable improvements in maternal health, child health, and child survival outcomes. In 2005, Chhattisgarh witnessed a sharp reduction by 11 points in the infant mortality rate, drawing considerable attention to this programme.

The core of the innovation is again institutional—developing and establishing norms that help select the right worker; providing a dedicated leadership; designing a capacity-building plan backed by human and financial resources; and devising appropriate monitoring strategies. There was also a need for internal advocacy to build respect for community processes amongst technical leadership and managers, and most importantly, to accommodate political interests and different stakeholder priorities. Such internal advocacy and championing the programme was possible because of a robust partnership between the government and civil society organizations active in community health. But perhaps what is most noteworthy is that regular innovation has become a feature of the programme. The list of such innovations is large: a welfare fund for mitanins, a mitanin-staffed help desk in government hospitals to assist patients referred by other CHWs to navigate the hospital and get appropriate care and counselling; literacy and equivalency programmes for mitanins to qualify to enter nursing schools; certification courses for mitanins; and so on. Such innovation has been possible because of the institutional framework, of which the most important is the creation of a state health resource centre, which has played the lead role in scaling up and supporting the programme and, along with the National ASHA Mentoring Group, has been promoting and facilitating further innovations in different states.

Establishing Quality Management Systems in Public Hospitals

Quality assurance has been a central concern of public health systems. Different states have tried different approaches, with little impact. In such a context, the NHSRC, the apex technical assistance institution that supported the implementation of the NRHM, has built up an innovative total quality management approach applicable to, and executable by, public hospitals. This approach touches upon clinical quality of care, administrative efficiency, patient amenities, and patient safety and satisfaction, and creates standard operating procedures for each of these. At the heart of the innovation is an understanding of what a quality management system is—an approach to seeing quality as a set of processes; identifying the current gaps in processes; coming to a technically sound and participatory consensus on how to close the gaps; documenting the processes and their improvements to meet the newly defined standard operating procedures; and building capacity to ensure that these are understood and adhered to.

Sixteen pilot hospitals developed the concept and evolved the tools, and then an effort to scale it up to 600 public health facilities helped understand the methodologies required for the expansion. Based on this experience and learning, the approach has gained approval as the quality policy framework for all public hospitals.

In contrast to other innovation pathways, this is an institution-led innovation. The NHSRC defines itself in its policy statement as aiming for a creative and innovative solution to address all the problems of public health services and has led similar innovations in health informatics and district planning. Its mandate, positioning, and management represent a significant institutional innovation.

***

Interestingly absent from the case studies discussed earlier are innovations related to information technology. No other technical domain pays so much effort and attention to innovation, and yet we chose to emphasize other, lesser-known and acknowledged forms of innovation in service delivery. Innovations based on the deployment of information and communication technology (ICT) are a mixed bag; most claims of success relate to putting a functioning information system in place, with the ability to display results. But when measured by the difference made to service delivery improvements or health outcomes, the evidence is less conclusive. The ICT innovations require a separate discussion and are therefore not included in this chapter.

Critical problem-solving is one of the most important and successful innovation drivers; to the extent that there is conscious learning from past experiences, there is always room to do better. Dedicated innovators—people who seek to innovate as an end in itself—seem to help; developments in technology, too, can stimulate or even constitute innovations, provided they help healthcare providers manage the 'institution–technology interface' better. This was workable when introducing rapid diagnostic kits or long-lasting insecticide-treated bed-nets for the control of malaria. In contrast, rapid and continuous innovation in health informatics has not been matched by noticeable, attributable outcomes in terms of improved service delivery or public health action.

We also note that, despite much attention to transformational or disruptive innovations, it is incremental innovations that have held sway. The 'Dial 108' service is a runaway success, as it supplements the public health system. The 'Dial 104' service, on the other hand, did not achieve similar success despite equal emphasis on its replication. The success rate of imported and donor-defined innovations—voucher models, for instance—is limited. Home-grown incremental innovations seem to do better.

Innovations flourish in certain ecosystems but not in others. Innovations take time. Institutional innovation needs effective decentralization, which primarily requires rules that allow considerable flexibility and space to create rules at decentralized levels. Improving public health services delivery needs an 'innovation mindset' that moves away from the 'enforcement mindset', which sees monitoring and accountability as the only available tools for improving service delivery. The latter approach assumes that the central problem of public sector delivery is that salaried workers have no incentives to do better. Innovations for public health services delivery also need better theory—going beyond making markets work for health to understanding and analysing institutional constraints, learning from past experience, and proposing alternatives.

Innovations have much to contribute to solving some of the most recalcitrant and intrinsic problems of public health services delivery. The most important step is in the mind—an attitude that sees innovation, not merely enforcement, as desirable and achievable—and that goes beyond the obvious to find solutions.

The drivers and sources of innovation are many and they are unpredictable, but only in a system that includes scientific validation and evaluation of claims can innovations bloom, flourish, and make a difference.

References

Ved, R., T. Sundararaman, G. Garima, and G. Rana. 2011. 'A Programme Evaluation of the Janani Suraksha Yojana', *BMC Proceedings*, 6(Suppl. 5): O15.

Sundararaman, T. 2012. *Opportunities, Ecosystems, and Roadmap to Innovations in Health Sector*. Sector Innovation Council for Health. Available at http://innovationcouncilarchive.nic.in/images/stories/sectoral/reports/Health_Sectoral-Report.pdf, last accessed on 14 March 2015.

8

VIJAY MAHAJAN

India as a Hub of Innovations for the Millions (I4M)

Innovations for the millions (I4M) are new products, processes, and institutional arrangements, often in combination with each other, which bring about a significant improvement in the quality of life of the masses—these days referred to as the 'base of the pyramid'—in a sustainable manner. In this chapter, 10 such innovations are described: three each from the private corporate sector, the government/public sector, and the non-governmental organization (NGO)/civil society sector; and one from the cooperative sector. By outlining the nature of these innovative efforts, their wide-ranging impact, both real and potential, on the well-being of the citizenry and the overall economic and technological progress of the nation can be seen. The purpose of this chapter is to introduce the idea that innovations that are relevant to the millions mostly do *not* come from the private corporate sector; thus, researchers need to cast their nets wider and look at other important actors in society.

Ten Innovations that Serve Millions of Indians

Private Sector

Bajaj Auto: Two- and Three-wheelers that Move Millions of Indians

Bajaj Auto Ltd is the world's third-largest manufacturer of motorcycles and the world's largest three-wheeler manufacturer. It fulfils the personal transportation needs of millions of Indians, as many of them migrate from rural India to its cities and face the prospect of commuting for the first time.

Venky's Poultry: From Hatcheries to Dressed Birds

Venkateshwara Hatcheries Pvt. Ltd was one of the pioneers in the Indian poultry industry. It established some of the first hatcheries in India, and then pioneered the concept of franchised hatchery operations. It also introduced and popularized cage-fed birds, as opposed to free-range. The company later established a chain of poultry stores by the name Venky's. Apart from improving the protein intake of millions of Indians, the company has generated self-employment opportunities for thousands of farmers.

Mobile Telephone Talk-time: One Cent an Hour!

The total number of active mobile telephone subscribers in India in April 2015 was 852 million, of whom 42 per cent were rural (*Trak. in* 2015). No other product or service in human history has had such penetration in mass markets. Though initially the call charges were high (as much as 40 US cents or USD 0.40[1] a minute), industry players such as Airtel dropped call charges to build up volumes and reap economies of scale. In certain telecom licencing regions, rates were as low as 1 US cent for an hour. The average revenue per user was less

---

[1] As per the exchange rate of USD 1 = INR 63.82, as on 31 May 2015.

than USD 3 per month. Yet, this is profitable for the telecom service provider and serves the communications needs of the millions. As much of India's workforce is in the informal sector, the ownership of mobile phones has significantly increased their productivity (Deloitte, GSMA, and Cisco 2012).

Government/Public Sector

Passenger Reservation System (PRS): Two Million Train Tickets a Day

Until the early 1990s, it often took several hours in queue to book a seat or berth in an Indian train. Indian Railways launched a pilot project for computerizing passenger reservations in 1985. The objective was to provide reserved accommodations on any train from any counter, prepare train charts, and account for the money collected. The full-scale system was launched in the early 1990s and has grown step by step. By 2014, Indian Railways was providing reservation services to nearly 1.5–2.2 million passengers a day on more than 2,500 trains running throughout the country.[2]

e-Governance Services: Millions of Indians Access Government Services through Information Technology Kiosks

India launched a National e-Governance Plan in 2003 to deliver government-to-citizen (G2C) services using Information Technology (IT) in order to improve access, enhance transparency, and reduce response time. These services, which include birth and death registrations, identity cards, driving licence and passport applications, utility bill payments, job applications, railway reservations, and more, are delivered through a network of over 125,000 IT kiosks called Common Service Centres, as well as by IT kiosk outlets run by state governments, such as APOnline. In 2016, 1.09 billion transactions had already been recorded.[3]

---

[2] See http://cris.org.in/CRIS/Projects/PRS, last accessed on 26 March 2017.

[3] See http://etaal.gov.in/etaal/YearlyChartIndex.aspx, last accessed on 26 March 2017.

Rashtriya Swasthya Bima Yojana

This case study has been covered in detail in Chapter 7. A national health insurance programme for the poor, Rashtriya Swasthya Bima Yojana (RSBY), was launched by the Ministry of Labour and Employment as a welfare scheme for millions of workers in the unorganized sectors. Every RSBY family pays only INR 30 for a biometric smartcard which entitles them to healthcare worth INR 30,000 per annum. The rest of the insurance premium is paid by the government. The insurance coverage is offered by private insurance companies on a competitive bidding basis, in collaboration with a network of 10,725 hospitals including 4,427 public and 6,927 private. As of 31 March 2016, 41,331,073 families had been enrolled and 11,841,283 had received hospital-based treatment.[4]

NGOs

Sulabh Sauchalaya: Enabling Millions in Urban India to Wash Up with Dignity

Globally, India continues to be the country with the highest number of people (597 million people) practising open defecation (World Health Organization 2014). But as far back as 1970, open defecation was becoming impossible in congested urban India. Public toilets were few and so badly maintained that no one could step near them. It was to address this problem that a young idealist, Bindeshwar Pathak, decided to set up serviced public toilets. He persuaded the municipalities of small towns and big cities, railways, bus transport companies, pilgrimage centres, and so on to give him land at prime locations to build attractive Sulabh toilet complexes, which had not only latrines but also bathing facilities. The complexes were maintained by workers from the erstwhile scavenger castes, now doing a much more dignified job. These were patronized by more than 10 million lower-income Indians every day,

---

[4] See http://www.rsby.gov.in/about_rsby.aspx, last accessed on 26 March 2017.

each paying as little as INR 1 (2 US cents) for use of the latrine and bath, along with a mini bar of soap.[5]

Aravind Eye Hospital: The World's Single-largest Cataract Operations Facility

This case study has been covered in detail in Chapter 7. Dr Govindappa Venkataswamy retired from a government job as an eye surgeon and decided to continue practising eye surgeries because millions of Indians suffered debilitating vision loss due to easily curable eye cataracts. Venkataswamy's sincerity inspired his family members and friends, and in 1976, he set up the Aravind Eye Hospital in Madurai, Tamil Nadu. The hospital offered the same high quality of care to every patient, rich or poor, paying or non-paying. A system was established to allow patients to choose whether they would pay a premium price, a cost-covering price, a subsidized price, or nothing at all. The latter group would still get completely free treatment along with free meals and stay. Hardly anyone misused this generosity; philanthropic contributions poured in to meet the deficit (Manikutty and Vora 2004). Costs were reduced due to volumes as well as systematizing operations. An intra-ocular lens factory was built, which reduced the cost of the lens from USD 80–100 to USD 5. Since its inception, Aravind has grown into a network of eye hospitals that have seen a total of nearly 32 million patients in 36 years and performed nearly 4 million eye surgeries.[6]

Microfinance: From NGO Microloans to Bank-led Financial Inclusion

Though India nationalized its major banks in 1969, the flow of credit to small borrowers never exceeded 20 per cent of the net bank credit. After 1992, when banking sector reforms were announced, this percentage steadily fell and several concerned NGOs working with the poor sought alternatives to bank credit. One of these was microcredit—loans as small as USD 100, usually given to women organized in a group who

---

[5] See http://www.sulabhinternational.org, last accessed on 26 March 2017.
[6] See Naidoo (2012).

guaranteed each other's repayment. The women used such loans for small-business activities like vegetable vending or raising small livestock for sale. Each family's increased income went partly to improving the family's standard of living and partly to repaying the loan. This simple idea grew from very limited access for the poor to credit from banks to nearly 100 million poor people getting credit worth USD 10 billion, most of it mobilized by microfinance institutions and NGOs from banks reluctant to lend to the poor directly (Microfinance Institutions Network [MFIN] 2014).

Cooperatives

Anand Milk Union Ltd (Amul): From Milk Powder Importer to the World's Largest Milk Producer

Amul was established in 1946 by Tribhuvandas K. Patel, a farmer leader who had the vision to hire a professional to take it to scale. Verghese Kurien, in 1949, then converted Amul into a vibrant dairy cooperative. After witnessing its success and understanding its growth potential, the then prime minister, Lal Bahadur Shastri, invited Kurien to replicate the model all over India, and then set up the National Dairy Development Board (NDDB) in 1965, appointing Kurien as its first chairman. Kurien propelled India's dairy development by using milk powder, received as a gift from European Community countries, the United States (US), and Canada, to manufacture reconstituted milk which it then sold. The proceeds were used to build a nationwide grid of millions of small milk producers organized into thousands of cooperatives, linked to hundreds of chilling plants, tens of processing plants, and thousands of vending points in distant urban markets, which sold to millions of consumers of liquid milk, milk powder, butter, yogurt, chocolate, and ice cream. By 2000, India had become the world's top milk-producing country, with 78 million tonnes produced annually (NDDB 2015).

Why Do All these Innovations Come from India?

Anybody visiting India does not fail to observe the paradox of India: though it is a rich country on the aggregate, with the third-largest gross domestic product (GDP) in purchasing power parity after China and the US, it also has the world's largest number of poor people—anywhere between 300 and 500 million, depending on the criteria used. The late Arjun Sengupta, an eminent Indian economist, asserted that 'a little more than three-fourths of the Indian people were poor and vulnerable in 2004–05' (Sengupta, Kannan, and Raveendran 2008). On the one hand, India has the sixth-largest number of billionaires, with a lifestyle and material comforts comparable to the best in the world; and on the other hand, one can see homeless pavement dwellers a few feet from the driveways of the rich. Rural poverty, which was desperate till the 1980s, has reduced thanks to overall growth and spending on rural employment programmes, but some of the rural poverty has spilled over to the urban areas through migrants, adding to the numbers of the endogenous urban poor.

Indian poverty is 'in your face'—a bit like the US big-city homeless—not tucked away in some remote rural areas, so that even the most insensitive person cannot miss it. So the Indian elite is always conscious of this fact. Different individuals respond differently to the same external circumstance. Some end up inured to poverty, while others make poverty alleviation a lifelong mission. India has a long tradition of having an 'elite of calling', a group of well-endowed, well-educated people who have worked to bring about greater justice, equality, and efficiency in society (Gupta 2013). This phenomenon dates back to the time of the freedom movement, which saw a number of individuals from elite backgrounds, including Mohandas Karamchand Gandhi and Jawaharlal Nehru, join the national cause.

In post-Independence India, some names which can be counted in this category include Dr Kurien of Amul and Aruna Roy, the tireless activist who virtually forced the central government to enact, first, the Right to Information Act, and then the National Rural Employment Guarantee Act (NREGA). Yet another phenomenon was that of individuals who grew up in modest or lower-income homes but due to education or entrepreneurship assumed influence in the corridors of power. This includes Lal Bahadur Shastri, India's prime minister after Nehru, who laid the foundations of both the Green Revolution in India, which made India self-sufficient in foodgrains, and the White Revolution (which converted India from an importer of milk powder to the world's largest milk-producing country), and Sam Pitroda, who ushered in the telecom revolution in India, which dramatically enhanced access to telephones without significantly increasing their numbers by setting up more than 2 million public telephone booths. Many of those working to reduce poverty do so because they feel that poverty and inequality lead to a waste of human potential.[7] Others believe poverty is a drag on growth that affects all of us. Many set up social enterprises, which are a major source of experimentation and incubation of I4M.

Special Characteristics of I4M

These 'I4M' are 'demographically appropriate'—they cater to millions, aiming eventually at the entire billion-plus population of India. They also recognize that a vast majority of users and consumers have limited purchasing power, as Sengupta underlines. These innovators have found ways to reach the three-quarters of Indians who are poor and vulnerable.

There are several ways by which this is achieved. One is cost-cutting: Aravind Eye Care System's Aurolabs, for example, manufactures world-standard intra-ocular lenses at extremely cheap rates. 'Micro-packaging' is another method. For example, shampoo, which poorer households cannot afford, were packed in biaxially oriented polypropylene single-use sachets priced at 10 US cents, which, while affordable to low-income customers, covered distribution margins and the additional cost of packaging, yet yielded a good profit to the manufacturer.

The I4M are rarely purely technical, but 'socio-technical' and often 'institutional', that is, they change the normative or contractual behaviour of players. One example of this is Sam Pitroda introducing the concept of payphone kiosks (called 'subscriber trunk dialling public call offices' or STD PCOs) in India in the mid-1980s. At that time, the

---

[7] Such young idealists first gathered under the banners of individuals like Bunker Roy of the Social Work and Research Centre, Tilonia, Anil Sadgopal of the Hoshangabad Science Teaching Programme of Kishore Bharti, and Dunu Roy of the Vidushak Karkhana, Shahdol; and later, organizations like Professional Assistance for Development Action (PRADAN) and the Bhartiya Agro Industries Foundation (BAIF).

telephone system was run by a government department, not a company, and the telephone ownership was less than 10 per 1,000 households. By establishing more than 2 million STD PCOs in a few years, Pitroda dramatically increased the availability of telephone services without significantly enhancing ownership. Through the Centre for Development of Telematics, Pitroda introduced technology upgrades and significantly improved rural connectivity by replacing old electromechanical exchanges, prone to breaking own due to heat and dust, with rural electronic exchanges, which were much more reliable.

Pitroda's real innovation was not just technical but 'socio-technical and institutional'. He realized that there was no chance of privatizing the telephone utility, so the only institutional reform he persuaded the government to do was privatizing the interface between the end user and the telephone service provider. Instead of surly departmental employees, ever ready to go home or on strike, the STD PCOs were operated by micro-franchisees of the telephone department who earned a commission on every call, and were thus ready to serve 24×7.

To ensure that charges were instantaneous and transparent, Pitroda's engineers also built a call metre, which flashed the call charges in bright light-emitting diode (LED) digits as the caller was talking. The STD PCO operator thus was user friendly and ready to provide service at any time of the day or night. These socio-technical and institutional dimensions of the innovation, coupled with the underlying technical innovation, effected important positive changes in the lives of millions of Indians who needed to make telephone calls, including long-distance calls. Had Pitroda followed the 'Washington Consensus' and argued for privatizing the telephone utility as a whole, it may not have worked in guaranteeing such widespread access to the utility at very affordable rates.

The I4M often arise in response to the 'failure' of the state. The overweening but increasingly ineffective state's self-appointed role as the alleviator of poverty and developer of economic potential leads to the government owning and operating a number of enterprises—from banks (over 70 per cent of India's banking assets are with state-owned banks)[8] and insurance companies (the Life Insurance Corporation [LIC]

---

[8] As on 31 March 2016. See Reserve Bank of India (RBI) (2016).

of India still accounts for 60 per cent of the business, 15 years after the government opened up the insurance sector and 19 international companies commenced operations in India)[9] to building national highways through the National Highway Authority of India (NHAI) and running hotels through the India Tourism Development Corporation (ITDC). These multifarious roles take away a lot of resources and talent from the more essential sectors of education, health, and poverty alleviation.

However, that does not prevent reform-oriented government officials from innovating as an 'inside job'. They work alone or in conjunction with private sector or NGO actors. While the government has resources and legitimacy, the private sector has much more flexibility in terms of deployment of resources, particularly for growth. It also brings in cost efficiency and a profit orientation. The NGO sector has insights to share regarding public welfare, often with deep contacts in the potential user group, which can be used for market research, as well as establishing new institutional arrangements. The three cases discussed earlier show all three types of reform effort: the state-owned Indian Railways PRS was entirely built by the railways' own Centre for Railway Information Systems. On the other hand, the RSBY uses private insurance companies to provide medical coverage through a competitive bidding process, with more than 8,500 public and private hospitals all over India providing the actual medical care. Finally, the growth of the Indian microfinance sector was initially based a lot on inputs from NGOs in designing its methodology.

The I4M often involve the use of state funding, either for early stage work, for capital expenditure, or for subsidizing ongoing operations. This is most true of government programmes, like the Andhra Pradesh Society for Elimination of Rural Poverty for women's empowerment and the RSBY health insurance. Some, like the Aravind Eye Hospital, rely on cross-subsidies, charging the rich more and the poor less but providing the same high-quality service to everyone. But not all models depend on state subsidy or cross-subsidy. Some have tried to come up with base-of-the-pyramid strategies which achieve market penetration using the customers themselves, such as Hindustan Unilever, which

[9] As on 31 March 2016. See https://www.ibef.org/industry/insurance-sector-india.aspx, last accessed on 26 March 2017.

uses women members of savings and credit 'self-help groups' as 'Shakti amma' rural retailers. Others have cut costs by moving procurement closer to the source of production. For example, the ITC Agri-Business Division uses village-level Internet kiosks (e-choupals) to procure agri-produce requirements straight from farmers at the well-known *mandi* (market yard) price, saving about 2 per cent on transportation and handling costs. Indeed, the diversity of financing innovations is worth a detailed study by business schools.

Constraints and Opportunities

Constraints

Despite growth in per capita incomes, a majority of the population still earns less than USD 2 per day, which constrains demand for many goods and services. The urban upper class and the urban and rural middle class are still relatively narrow, accounting for no more than 25 per cent of households. Corporations have tried to address this constraint in a number of ways, such as financing purchases. This was done in a big way for two-wheelers and cars and certainly expanded the demand base. Housing is another area where this has been attempted with success, mostly in the middle-class market. Another way in which the demand base has been expanded is through micro-packaging at the single-use level. Multiple single uses, whether of hair shampoo or mobile talk-time, then add up to a large demand base. The expanded base then leads to economies of scale; if part of this is passed on to consumers, costs come down and demand goes up, leading to the virtuous cycle we saw in operation in the Indian mobile telephony sector.

The government's attitude towards I4M has generally been negative at worst, or stand-offish at best. Government officials rarely embrace any innovation, as it always leads to changing existing ways and patterns, which for many have become a habit and for many others entrench vested interests. Other innovations require early stage risk-taking and potential failure—anathemas to a bureaucracy which is generally penalized only for failed attempts and not inaction. This leads to a 'play safe, avoid change' attitude. Many innovations require legislative and regulatory changes; this itself is treated as sufficient reason not to adopt them. This is the reason that not many successful innovations emanate from within the government or the wider public sector and are led by bureaucrats.

Society's attitude towards I4M has been more positive, but Indian society still does not perceive commercialization positively. There is a widely held view that the poor and lower-income groups should be served at subsidized rates or even for free. Anyone seen to cover costs is already suspect; and someone who profits by serving the poor is easily branded an exploiter. This attitude culminates in the Indian Income Tax Act of 1961, which provides that if a not-for-profit organization earns fees from services (beyond the limit of USD 40,000 per annum), then it is taxed as if it were a business. This forces legitimate actors to remain subsidy or grant dependent, whereas others learn to circumvent the law.

To summarize, unless there is wider societal consensus that it is legitimate to make a reasonable profit by charging fees for services or prices for products aimed at the base of the pyramid, and this is reflected in government policies, tax legislation, and regulation, there is little scope for ongoing efforts for I4M to become sustainable.

Most research and development (R&D) labs are owned and funded by the government. Labs in universities and the Indian Institutes of Technology (IITs) are again publicly funded, though now they are diversifying their sources of support. There are few corporate and NGO R&D centres. Though one would imagine that state-funded labs would have a bias towards the under-served, the experience has been of disconnection with markets and pumping resources into the favoured ideas of hierarchical senior scientists and bureaucrats. As a result, most government-funded labs have hundreds of patents and prototypes for things that attract no demand.

Only when a visionary leader takes over do these behemoths turn their attention to 'relevant problems'. For example, Vikram Sarabhai was able to give the Indian Space Research Organisation (ISRO) a developmental mission and focus on applications such as agro-meteorology, remote sensing of earth resources, and satellite instructional television, all of which eventually served millions. In the private sector, as chairperson of Hindustan Lever, T. Thomas managed to get his R&D lab to focus on substituting edible oils (which were imported using scarce foreign exchange) with sal seed oil for soap manufacture, which not only saved edible oils but also created income-earning opportunities for tribal people who collected sal seeds from forests.

The R&D is often focused on the technical, whereas the issues are often socio-technical and even institutional. This leads to many promising innovations not being taken up at scale.

Opportunities

India's large population base, mostly young and with increasing aspirations and incomes, is a dream for an innovator: demand is assured. Coupled with the history of shortages or indifferent provision of goods and services by state or private monopolies, demand has been so bottled up that it startles even the most optimistic marketer. For example, when LML Vespa, then a new manufacturer for two-wheeler scooters, opened 'bookings' for scooters in 1979, it got enough bookings to create a waiting list 21 years long, based on annual production at that point in time. This also explains the exponential jump in the production of the Bajaj two-wheelers, noted earlier.

Many innovators have assumed iconic status. Examples are Amul's Verghese Kurien, Bajaj Auto's Rahul Bajaj, Telecom's Sam Pitroda, mobile telephone operator Airtel's Sunil Bharti Mittal, Sulabh's Bindeshwar Pathak, and Aravind's Govindappa Venkataswamy, all of whom inspire thousands of youngsters to address similar problems. This definitely creates a supply of human talent for the next round of innovations. In addition, a vast population of diaspora Indians, many in advanced professions, who bring experience of how things work—in terms of technical know-how, financing, and institutional arrangements—are either returning to India for good or investing in India. In both cases, they bring valuable experience of successfully working in other cultures and a degree of impatience with things being stuck in India. This energizes the entire ecosystem for incubating I4M.

Commercialization opportunities, too, are rising. Seed funds, angel investors, and early-stage venture capital funds are all beginning to come up, though nowhere near what is needed. Funding, particularly for the early stage, is increasingly coming in. Inter-sectoral collaboration is also increasing. The process of trust-building that leads to useful collaboration amongst governmental, private, and NGO actors is beginning to gather momentum. This will unleash a huge potential for I4M.

Fostering an Innovation-supportive Ecosystem

In order to foster an ecosystem which is supportive of I4M, we need to take a number of steps, but the most crucial is to attract bright young people who not only have the empathy with the masses but also come up with alternatives to this field. This requires, in other words, coupling the inquisitive mentality of a social science or hard science student with the problem-solving approach of an engineering or management student.

There are now a number of places where such supportive structures have come up. A notable example is the Indian Institute of Management Ahmedabad (IIMA), where Anil Gupta first established Srishti, an NGO for I4M, and then collaborated with the state government of Gujarat to establish the Grassroots Innovation Action Network and later, with the government of India, to establish the National Innovation Foundation. Another group of professors at the IIMA has promoted the Centre for Innovation, Incubation and Entrepreneurship, which in turn has helped incubate several social enterprises.

All the original five IITs—in Kharagpur, Mumbai, Kanpur, Chennai, and Delhi—have incubators established with funding from the government's Department of Science and Technology, which provides grants as well as equity support to technology entrepreneurs.

Another example of incubation is the Aavishkaar India Micro Venture Capital Fund, which organizes an annual event called Sankalp, attracting hundreds of entries; investors as well as potential entrepreneurs throng there to find a match. Intellecap, a sister company of Aavishkaar, conducted a survey which found that USD 1.6 billion of capital had been invested in 220+ 'impact enterprises' (a term increasingly used for enterprises trying to serve the base of the pyramid population segment) across India (Intellecap 2014). Funding, particularly for the early stage, is increasingly forthcoming, from non-profit funds like Dasra and UnLtd as well as angel investor networks and formally incorporated venture capital firms like Aavishkaar, Unitus, Elevar, and Sequoia. The latter was a standard private equity investor until a few years ago, but has now raised a USD 500 million fund for early stage investments. Support entities like design firms (such as Ideo in Bangalore and Instillmotion in Hyderabad) and fundraising advisers (such as Intellecap and Unitus Advisors) have also arisen. Business plan competitions like the one at Sankalp, the Global Social Venture Competition at the Indian School of Business in Hyderabad, and a number of events conducted by The Indus Entrepreneurs network (itself incubated by successful Silicon Valley entrepreneurs of Indian and South Asian origin), from start-up boot camps to 'how to make a pitch' sessions, put potential entrepreneurs face to face with angel investors. A supportive ecosystem for I4M is emerging fast in India.

***

Not only are these I4M improving the quality of life of millions of Indians, they are doing so while saving government expenditure and releasing the government's energy to perform more critical governance functions. They also create opportunities for the private corporate sector to scale up in emerging markets. These innovations are equally useful for most other developing countries, so multinationals like Lucent, Philips, BP, Shell, Unilever, Essilor, Schneider, and Panasonic have all set up their global hubs for base-of-the-pyramid product development in India (*Business Standard* 2012; Gibson 2013). Schneider Electric (n.d.) began exporting its In-Diya brand of low-cost solar home-lighting systems to several African and Latin American countries in 2013.

Several I4M have reached out to the base-of-the-pyramid markets in Africa and Southeast Asia. The Bajaj three-wheeler can be seen in many parts of Africa and is becoming a popular low-cost means of short-distance transport for people and goods. The cost competitiveness of mobile telephony in India, necessitated and propelled by reaching out to the base of the pyramid, has enabled Bharti Airtel to spread to several African countries. Indepay, a low-ticket payment solutions company working with the State Bank of India, has struck a deal to create a nationwide network for payments with the largest private bank in Indonesia. BASIX, the global pioneer in devising weather index-based crop microinsurance, has provided advisory services to launch weather-index-based insurance in Malawi and Zambia (United Nations Framework Convention on Climate Change n.d.).

The Government of India has committed over USD 11 billion in international aid, mostly in the form of lines of credit to developing countries, to help import goods and services from India. Over half of this has been committed to Africa. While a lot will go into big infrastructure like railways, some Indian I4M are also included, such as IT kiosks to provide public services to citizens.

Many Indian NGOs and civil society organizations have begun reaching out to the base-of-the-pyramid markets in Africa and Southeast Asia. Sulabh International started work to streamline sanitation facilities in Ethiopia, Angola, Madagascar, Tajikistan, and Haiti in 2009. The CAP Foundation established vocational training centres in Kenya; the Naandi Foundation has established Araku as an international brand for organically grown coffee and is now sourcing beans from Uganda and Mozambique in addition to the original Araku Valley in Andhra Pradesh.

These are early signs of India becoming a global hub for I4M. These innovations are different from the ones that emerge from Silicon Valley and have the potential for serving billions at the base of the pyramid. As India becomes the fastest-growing large economy in the world, it may find that its most valuable exports are the products, services, and institutional arrangements developed by the Indian I4M.

References

Business Standard. 2012. 'Panasonic India Sets Sight on Bottom of the Pyramid', 19 April. Available at http://www.business-standard.com/article/companies/panasonic-india-sets-sight-on-bottom-of-the-pyramid-112041900040_1.html, last accessed on 26 March 2017.

Deloitte, GSMA, and Cisco. 2012. 'What is the Impact of Mobile Telephony on Economic Growth?'. Available at http://www.gsma.com/publicpolicy/wp-content/uploads/2012/11/gsma-deloitte-impact-mobile-telephony-economic-growth.pdf, last accessed on 26 March 2017.

Gibson, A. 2013. 'Lighting the Path of Profitability for Base-of-the-Pyramid Enterprises', *Devex*, 18 October. Available at https://www.devex.com/

news/lighting-the-path-of-profitability-for-base-of-the-pyramid-enter-prises-82126, last accessed on 26 March 2017.

Gupta, D. 2013. *Revolution from above: India's Future and the Citizen Elite.* Bengaluru: Rupa.

Intellecap. 2014. *Invest. Catalyze. Mainstream: The Indian Impact Investing Story.* Mumbai: Intellecap. Available at http://www.intellecap.com/publications/invest-catalyze-mainstream-india-impact-investing-story, last accessed on 26 March 2017.

Manikutty, S. and N. Vohra. 2004. 'Aravind Eye Care System: Giving them the Most Precious Gift', Indian Institute of Management Ahmedabad. Available at https://wiki.brown.edu/confluence/download/attachments/9994241/aravind+case.pdf?version=1, last accessed on 26 March 2017.

Microfinance Institutions Network (MFIN). 2014. 'Micrometer: Statistics on Microfinance in India', Slideshow, December. Available at http://mfinindia.org/wp-content/uploads/2014/06/Micrometer%20Issue%2012_Q3%20FY%2014-15_12th%20Feb%202015.pdf, last accessed on 26 March 2017.

Naidoo, Jayaseelan. 2012. 'An Infinite Vision: The Story of Aravind Eye Hospital', Huffington Post, 15 May. Available at http://www.huffington-post.com/entry/an-infinite-vision-the-st_b_1511540.html, last accessed on 26 March 2017.

National Dairy Development Board (NDDB). 2015. 'Genesis'. Available at http://www.nddb.org/English/Genesis/Pages/Genesis.aspx, last accessed on 26 March 2017.

Reserve Bank of India (RBI). 2016. *Report on Trend and Progress of Banking in India 2015–16.* Mumbai: RBI.

Schneider Electric. n.d. 'Offers for the Base of the Pyramid'. Available at http://www2.schneider-electric.com/sites/corporate/en/group/sustainable-development-and-foundation/access-to-energy/solutions.page, last accessed on 26 March 2017.

Sengupta, A., K.P. Kannan, and G. Raveendran. 2008. 'India's Common People: Who are they, How Many are they and How Do they Live?', *Economic and Political Weekly*, 43(11): 49–64.

Trak.in. 2015. 'Indian Mobile Subscriber Stats', 11 April. Available at http://trak.in/tags/business/2015/04/11/indian-mobile-subscriber-growth-num-bers/, last accessed on 26 March 2017.

Venkateshwara Hatcheries. 2010. 'Group Profile'. Available at http://www.ven-kys.com/group-profile/venkateshwara-hatcheries-pvt-ltd/, last accessed on 26 March 2017.

United Nations Framework Convention on Climate Change. n.d. 'Micro-insurance Reducing Farmers' Exposure to Weather Risk'. Available at https://unfccc.int/files/adaptation/application/pdf/basix.pdf, last accessed on 26 March 2017.

World Health Organization. 2014. *Progress on Drinking Water and Sanitation*. Available at http://www.who.int/water_sanitation_health/publications/2014/jmp-report/en/, last accessed on 26 March 2017.

9

BRIAN ENGLISH[*]

Market-Based Solutions for Poverty Reduction in India

India's rapid urbanization has been marked by a significant increase in urban slum populations, with official estimates projecting it to climb above 100 million in 2017 (Ministry of Housing and Urban Poverty Alleviation [MHUPA] 2013). Urbanization and the urbanization of poverty poses serious and pervasive challenges to the country and demands innovation across many sectors to help India reach its full potential. India has been a beacon of innovation in market-based solutions in poverty reduction, creating enterprises that span from sanitation to education improvements. This chapter profiles two case studies of market-based solutions in India that provide some important lessons for others fostering similar initiatives: (a) LabourNet, a social enterprise that provides vocational skills for informal-sector workers, which began in Bengaluru in 2007 and has now scaled up to 13 regions of India; and (b) The Trash to Treasure programme, which started as a pilot to test enterprise models for recycling waste in Bengaluru and has evolved into

* The original research for the chapter was conducted by the author in India between 2009 and 2012, funded by the Bill and Melinda Gates Foundation. It included a longitudinal process evaluation and impact evaluation.

a network of organizations that are establishing one of India's first city-wide recycling programmes. Both cases demonstrate how partnerships between local government and social enterprises can deliver large-scale results. Both cases also show how important the enabling environment is to these supply-driven enterprises, in addition to fine-tuning their business models.

Rethinking Assumptions about Slums, Poverty, and Solutions

As urbanization has spread rapidly over the past two decades in India, so has there been a rise in poverty and slum conditions in the cities.[1] As of 2017, an estimated 100 million people live in urban slums in India, which is about 25 per cent of the country's urban population (MHUPA 2013). As the urban population swells to an estimated 590 million by 2030, the slum population is expected to grow faster than overall urbanization rates in many cities, particularly if the provision of affordable housing and land for the poor does not keep pace with the demand (Sankhe et al. 2010).

Urbanization in India is not a side effect of its economic growth, but rather an integral part of this process. For centuries, cities have proven themselves as engines of economic growth, sources of innovation, and places of job creation. India's cities have expanded rapidly as increasing numbers migrate to cities in search of social and economic progress. Today, India's urban economies make up the major portion of the country's economy, generating two-thirds of the national gross domestic product (GDP) and 90 per cent of government revenues.

While India's high urban growth rates are an indicator of success, the rise of slums in India, and elsewhere in the world, reflect a number of failures—of governing institutions, dysfunctional markets, lack of political will, and policies and practices of exclusion. The approaches to improve and prevent slums in India have progressed substantially over the past few decades as local and national governments began

[1] India's urban population grew from 217 million in 1991 to 377 million in 2011 (MHUPA 2013).

to recognize that slums are not going to vanish on their own. Despite the increased attention and action by the government at all levels, the growth of slums has far outpaced the impact of interventions undertaken to date (Sankhe et al. 2010).

Although urbanization is increasingly concentrating the spatial location of poverty in India today, it also provides the possibility to escape poverty (Gupta et al. 2014). This very search for greater social and economic progress drives migration to cities, contributing to the growth of slums. Slums are where the city's informal workers go to sleep. However, outsiders have many misperceptions about the urban poor and slum communities. They view them as destitute, incapable, or backward. Slum communities are not static settlements, just as slum dwellers are not 'stuck'.

Slum residents are one of the more resilient populations within cities and have developed sophisticated coping mechanisms to deal with regular shocks and volatility. Slum residents have long demonstrated that they can mobilize their own resources and networks, and save and invest in the betterment of their communities, when given the chance. However, when top-down decision-making treats slum residents as welfare recipients, this underestimates their capacities and reinforces patronizing relationships. Instead, successful programmes need to deliver tools to the poor that help them gain skills and connection with the wider economy, as well as ownership, education, and security. They need to engage residents as active agents of change and seek to unlock the economic potential of these communities.

Unlocking the Economic Potential of Slums and Delivering Opportunities

The complexity of urban systems and poverty prevents any 'one-size-fits-all' solutions. Indeed, solutions require all sectors to work together. The public sector has enormous power to create policies, direct resource flows, and design instruments that can complement and encourage civil society's and private sector's engagement in the slums. Civil society, with its understanding of the issues of poverty, its empathy, its passion, and its ability to give voice to the poor, has a critical role to play in creating solutions that work. The private sector, with its creativity, boldness, and penchant for innovation, also has great potential to engage constructively in seeking new ways to unlock the economic potential of the urban slums and deliver missing goods and services.

Market-based solutions have been gaining the attention of governments, international aid groups, NGOs, and entrepreneurs as they look for new ways to scale up their impact and sustain their interventions. The promise shown by the microfinance revolution has also given rise to aspirations by entrepreneurs, investors, and NGOs who seek to harness market forces for mission-driven, long-term development goals. Many of these organizations are viewing the poor's unmet needs as untapped market opportunities. This has given rise to a wave of new 'social enterprises', 'inclusive business models', and 'impact investing'.

The book, *The Fortune at the Bottom of the Pyramid* (Prahalad and Fruehauf 2004), perhaps set off the first wave of investigations into how low-income markets could present opportunities for companies seeking new fortunes, while also bringing prosperity to the aspiring poor. Since then, many global companies have been able to operationalize profitable models in places in India and elsewhere, such as Unilever and Cemex. The predominant focus of these pursuits has been on fine-tuning business strategies at the base of the pyramid.

The global financial crisis, new technologies, and an emerging class of socially conscious entrepreneurs have spurred another wave of social enterprises and market-based development programmes, all looking to scale up their impact and sustain their solutions with revenues rather than depend on grants. Governments and donors with tighter budgets are supporting such initiatives and a host of development organizations—from non-profits to private sectors—are incorporating such approaches into their portfolios.

Indian Case Studies

Across all sectors in India, entrepreneurs and traditional development organizations are harnessing market-based solutions and creating new enterprises to address some of the toughest problems in education, energy, health, housing, water, and finance. Impact investment funds have also developed, like Acumen, to invest in these companies, leaders, and ideas. Donors, such as the Bill and Melinda Gates Foundation, are supporting cities in India to test new sanitation solutions for the urban poor by engaging the private sector, rather than solely pursuing the public provision of traditional water- and energy-intensive systems.

Global Communities, an international NGO working in India since 2003, has been incorporating new market-based approaches to address urban development challenges for the poor, especially those living in slums. Through a programme called Slum Communities Achieving Liveable Environments with Urban Partners (SCALE-UP), Global Communities supports two particular initiatives that represent two ends of the spectrum of solutions being tested: (a) LabourNet and (b) the Trash to Treasure programme.

SCALE-UP was predicated on historic experiences which show that the responsibility for bridging the 'convergence gap' between governments, various stakeholders, and slum communities cannot be left solely to either local governments or slum residents. Successful approaches often come from the middle, from intermediary institutions—NGOs, civil society groups, even private enterprises—that can engage communities and local governments in productive solutions by filling gaps in information, trust, and technical skills to implement projects. The two examples given next illustrate this.

Case Study: LabourNet

Throughout the slums of India, employment and income generation are paramount to the ability of the urban poor to raise themselves out of poverty. One great obstacle is the ability of informal labourers to integrate into the formal sector, earn higher wages, find stable employment, and gain social benefits such as health insurance. LabourNet is a social enterprise that helps to integrate the urban poor in India's construction sector by improving the skills of labour and small enterprises to respond to sector demand and move upward into higher-value markets.

The construction sector represents a substantial sector of employment for India's urban poor, employing more than 45 million people and allowing for large-scale absorption of rural labour and unskilled workers, in addition to semi-skilled and skilled workers (National Skill Development Corporation [NSDC] 2013). It provides opportunities for seasonal employment, thereby supplementing workers' income from farming, and permits large-scale participation of women workers. Almost 3.5 per cent of urban micro and small enterprises (MSEs) in India are engaged in the construction sector. This number increases significantly when accounting for the additional 45.4 per cent of MSEs that are engaged in trading and repairing services, a significant portion of which encompass construction-related repair and maintenance (Sharma and Chitkara 2006).

The industry is integral to the overall economy of India. The construction sector was estimated to contribute about 8 per cent to the country's GDP between 2012 and 2017 (NSDC 2013). In addition, construction accounts for nearly 65 per cent of the total investment in infrastructure and is expected to be the biggest beneficiary of the surge in infrastructure investment over the next five years (2013). This industry offers significant economic multipliers due to the number of upstream and downstream markets in construction value chains.

The urban poor are integrated at the lower end of the different value chains within the construction sector, which is marked by widespread informality. These workers can be divided into two broad categories: labourers and workers as part of MSEs. Labourers are typically integrated into the production chain through subcontractors—labour contractors employed by larger companies. Subcontractors, while utilizing the same access points as individual labourers, also secure additional business directly from clients like single-family homeowners and small business/office project owners, provided they are successful at networking and negotiating business deals.

Most labourers enter the industry via informal networks, usually recruited from rural and peri-urban areas by a subcontractor who has a connection to their village. Several of these labourers work as unskilled helpers, migrating seasonally to work on construction sites and returning to their homes several times annually for farm work and holidays. As unskilled helpers, they may work across several trades, from painting to masonry to bar-bending, carrying tools, moving supplies, and other manual labour, with a wage typically ranging from INR 70 (USD 1.16) up to INR 100 (USD 1.66) per day depending on the project, as well as, often, less transparent factors such as gender and age. Labourers who work in the industry for a year or more and have aptitude may then find an opportunity to become semi-skilled or skilled in a specific trade. This may enable them to move up to salaries ranging from INR 150 (USD 2.50) to INR 300 (USD 5.00) daily.

There is, however, no defined path to becoming a skilled labourer; the status of individual labourers may vary based on the nature of the job and from project to project, since there are no formal means of establishing a worker's skill level and years of experience. As a result, labourers have to negotiate anew for each project they undertake. A labourer's path to becoming skilled may also be closely tied to the sub-contractor with whom they work. So, in effect, the subcontractor also has a great deal of say in a labourer's career path, pay, and experience. This system of recruitment and on-the-job training leaves labourers open to exploitation.

Integrating the Urban Poor into the Construction Sector

Employers, industry, and government across India regularly lament the lack of qualified workforce—at all skill levels—as the key constraint to the industry's ongoing growth. Skill development is therefore the primary opportunity for India's urban poor to upgrade their livelihoods and increase their incomes. On the supply side, there are few training opportunities accessible to the urban poor. Few people have information on where to get training, while the means to access it (such as transport, location, and timings) and the cost often create barriers for the labourers.

In 2007, an NGO called Movement for Alternatives and Youth Awareness, which worked largely with child labourers and was acutely aware of the obstacles facing migrant labourers coming to cities, realized that they could play a stronger role in supporting the integration of migrant and informal labourers into the market by becoming a force in the market themselves. They developed LabourNet, with a business model to fill the gap between supply and demand of trained workers and offer social benefits from which informal labourers were excluded.

As Bengaluru's economy boomed, LabourNet saw an opportunity to train and link workers with jobs amongst the growing middle-class, mid-sized business community and the active construction sector, which required reliable and 'skilled' workers. The potential employers were facing difficulties finding and training the requisite labour force, thereby turning to placement agencies like LabourNet (which specifically caters to the construction industry). LabourNet sought to provide informal construction and service sector workers with institutionalized access to jobs, enhanced incomes, and financial and social services.

Workers register with LabourNet by paying a small membership fee, which provides them with job placement, training, health insurance, an identification card, and a no-frills bank account at Punjab National Bank. LabourNet receives service requests through call centres and a Web-based interface, and then broadcasts these job openings to qualified workers. LabourNet uses mobile phones and SMS messages to announce job opportunities and provide service providers with the opportunity to bid on available jobs.

After about four years of fine-tuning, LabourNet's leaders concluded that the scale of workers they were able to place at jobs did not meet their objective of assisting hundreds of thousands of workers. Nor did the profitability of this model lend itself to long-term sustainability. LabourNet, therefore, recalibrated its focus to develop business relationships with large business organizations that had significant and recurring demands for qualified construction workers.

LabourNet customizes its construction content for each client provides training through dozens of sites across the country. Where workers have poor skill levels and limited training access, LabourNet is engaged to conduct training for both supervisors and workers, as well as post-training assessments.

The success of LabourNet's business model can be seen in how quickly it has scaled up. Between 2007 and 2011, it trained more than 43,000 informal workers, provided accident insurance to 32,000, opened bank accounts for 14,000, and linked more than 8,000 workers with jobs. From 2011 to 2013, LabourNet scaled up its operations to 23 livelihood centres, 71 schools, and 185 on-site training camps across 25 states of India. By 2013, LabourNet had trained 100,000 workers. They have now set their aim to improve the real income of 20 million individuals in the informal sector by 2020.

Case Study: Trash to Treasure

Bengaluru's population of almost 8 million people produces about 3,500 tonnes of waste daily (Karnataka State Pollution Control Board [KSPCB] 2013). The systems that manage waste in the city encompass a complex set of formal and informal enterprises and relationships between diverse actors who collect, recycle, and dispose of waste every day. Around these systems converges a nexus of issues regarding poverty, government services, enterprise development, and environment protection. The challenges that Bengaluru faces are indicative of what most cities across India are facing, or will be soon facing, as the country urbanizes. India's Ministry of Urban Development (MoUD), in its 2016 manual on solid waste management, estimates that India generates nearly 150 million tonnes of waste per year, of which over 40 per cent of this waste is never collected in the first place or properly handled after collection.

There are as many as 1.5 million informal waste pickers in India—mostly women and the majority from lower castes—who retrieve recyclable materials from city streets and dumps to support their families. The recyclable materials are sorted, typically in back alleys or vacant lots, and sold for a small profit to mid-level junk dealers, who then sell to wholesalers, who bundle large quantities of the recyclable waste and sell it onward to corporate recycling plants. On average, a self-employed waste picker earns about INR 100–200 (USD 1.75 to USD 3) a day, based on a 2010 survey in Bengaluru (CHF International and Mythri Sarva Seva Samithi 2010). This informal sector forms the backbone of the nation's nascent recycling economy. It is estimated that waste pickers collect more than 10,000 tonnes of reusable waste every day across the country. Annually, this generates some USD 280 million in revenue (Kapur 2011).

The occupational health hazards are numerous, including harassment from the police, who treat waste pickers as 'thieves'. Yet, the service they provide to municipal waste management systems and the environment is largely ignored. Their recovery and sale of recyclable materials creates 'green jobs' downstream in the processing of these materials and reduces waste going to landfills, free of cost to the city, along with reductions in greenhouse gas emissions (KSPCB 2013).

In the mid-1990s, the Indian Supreme Court ruled that a failure to collect and handle municipal waste was a violation of citizens' fundamental rights. Its decision prompted the government to propose a new set of municipal solid waste collection rules in 2000. Waste pickers, amongst the nation's poorest residents, were not afforded a role in collection and segregation under these rules. Instead, many cities contracted with corporate waste collection companies to take over the responsibility for waste collection at bin sites—and in some cases, from homes—and transport it to landfills. These companies are paid by the tonne for the trash, so they maximize collection but do not segregate and recycle. By taking over municipal bins and encroaching on door-to-door collection, the companies have reduced waste pickers' access to municipal waste, endangered the livelihoods of the poor, and put economic pressure on waste pickers to pursue the more risky alternative of collecting at landfill sites.

Global Communities, in partnership with a broad group of supporting organizations and civil society groups, created a programme called Trash to Treasure in 2008 to address both recycling in the city and the plight of marginalized waste collectors. The programme has been piloting and promoting neighbourhood recycling centres where waste can be segregated, stored, and sold to wholesale recycling markets. The recycling centres give waste pickers legitimate space to work, so they can collect and sell more material and make a bigger wage, free of harassment by locals, and with better working conditions. The business model for these recycling centres is simple: the city provides the land and building costs, Global Communities uses grant money to kick-start the operations of the centres, and local NGOs or social enterprises manage the operations of each centre. They hire and pay the waste pickers through two fees: a monthly fee from households for collecting waste and a fee from larger recyclers that sell recyclables and organic waste. The economic viability of these units was piloted during the first two to three years of the programme, with funding from the Bill and Melinda Gates Foundation. Now, Global Communities has leveraged additional funding from the Caterpillar Foundation and the municipal government of Bengaluru to open a total of seven recycling centres across the city.

Based on the success of this programme, Global Communities has been helping the municipal government of Bengaluru to roll-out and implement a larger 'decentralized' waste management system, with the city building one recycling centre in each of the 198 wards. The system aims to increase the amount of waste being recycled, create new jobs and incomes for workers at recycling centres, and integrate informal sector workers into these activities. The programme catalyses the talent and energy of both entrepreneurs and small enterprises that generate income and revenues from selling recyclable waste and service fees. The seven waste recycling centres that the Trash to Treasure programme created or supported in Bangalore have the capacity to recycle 50 tonnes of waste per month.

A Roadmap for Integration: Legitimizing Informal Waste Collectors in the City

Over the last few decades, there has been a slow but gradual recognition of the contribution of waste pickers in both Indian law and policy. Noteworthy amongst these are Child Labour (Prohibition and Regulation) Act, 1986; the Municipal Solid Waste (Management and Handing) Rules, 2000; National Environment Policy, 2006; National Action Plan for Climate Change, 2007; and Unorganised Workers' Social Security Act, 2008. More recently, the Plastic Waste Management and Handling Rules, 2011, and the Draft Rules of Municipal Solid Waste, 2011, mandated that urban local bodies engage waste pickers in the collection of plastic and other waste. However, these rules do not provide details about implementation.

To provide an impetus for Bengaluru's local government to implement these policies, Global Communities helped submit an affidavit in a Lok Adalat—a 'people's court' for alternative dispute resolution—which provided a roadmap for integrating waste collectors into the city systems and protecting their safety while rewarding their contributions. Global Communities leveraged this directive and helped the city define its processes for enumerating informal waste pickers—estimated at 20,000 across the city—and also supporting the formation of an association representing about 2,500 waste pickers in the city, called Hasirudala, 'green force', which advocates for better working conditions and access to additional government services, such as healthcare. In tandem with this, Bengaluru's local government also became the first city in India to issue identity cards to waste pickers that authorize their work in the city. Over 7,000 informal waste collectors have received these cards (Chakraberty 2014).

These are significant steps forward in supporting the rights, dignity, recognition, and security of waste pickers. It helps legitimize these workers in the city and ultimately integrate them into formal waste management systems. Through these actions, Bruhat Bengaluru Mahanagara Palike (BBMP) has demonstrated leadership in the country and provided a model for other cities.

Navigating the Waste Web: Harnessing the Informal Sector in Formal Solutions

Studies of this sector by Global Communities and others have continually highlighted that the Indian formal sector is supported by an extremely robust informal sector. However, most of the formal-sector waste management solutions that cities across India are employing focus on collecting and shipping waste to landfills, without any formal recycling programmes. This model overlooks the existing recycling in the informal sector.

Experience shows that it can be highly counterproductive to establish new formal recycling systems without taking into account the informal systems that already exist. The more effective option is to integrate the informal sector into waste management systems, building on their practices and experience, while working to improve efficiency and the living and working conditions of those involved. The decentralized waste management approach that Global Communities has been promoting and demonstrating provides a tremendous opportunity for integrating the informal sector, which consists of individuals who work with waste materials on a daily basis and are essentially experts on material identification, value, and recovery.

When the System Broke Down

Bengaluru's municipal corporation, BBMP, has been dumping the city's garbage in the landfill of the neighbouring village, Mavallipura, for years. This dump is managed by a private company contracted by the BBMP. In September 2012, the Karnataka Pollution Control Board closed down the landfill after the villagers staged opposition to the unscientifically managed landfill that was polluting their environment (*The Times of India* 2014). Villagers did not allow BBMP trucks to enter the landfill near their neighbourhood. For more than 10 days, garbage was left uncollected in the city and began piling up in neighbourhoods.

Troubled by this crisis, BBMP and other stakeholders, including media, turned to Global Communities and its partners to advise on new systems for the city on recycling and decentralized management of the municipal waste. This provided additional impetus for the city to continue expanding the decentralized recycling centres throughout the city and for residents to begin segregating recyclable materials before giving them to waste collectors. Informal waste collectors have a tacit knowledge about materials and processes and this immense knowledge pool of the informal sector is largely unacknowledged. Global Communities and many other organizations have been trying to highlight their value, validate their role, and help them be seen as 'skilled labour' or 'knowledge workers', rather than merely 'labour'.

Various studies have also documented the waste pickers' contribution to the reduction of municipal waste-handling costs, resource recovery, environmental conservation, and climate change mitigation. The most notable study was completed by the Expert Committee on Solid Waste Management constituted by the Hon. Supreme Court of India. Even India's 2008 National Action Plan on Climate Change has lauded the informal sector as the backbone of India's recycling system and affirmed its role in emissions abatement.

These projects supported by Global Communities have found that formal and informal sector workers suffer not only due to the collective apathy around solid waste management and the manner in which the public views waste pickers, but also in the way they view themselves. Nobel Prize-winning economist George Akerlof, in his book co-authored with fellow economist Rachel Kranton (2010: 151), explains how people's identity—both their own perceptions and their societies' perceptions—is one of the most important influences on their economic well-being. Waste collectors in India suffer from social norms associated with lower caste and an identity that is marginalized, along with how society views solid waste, pushing them 'out of sight and out of mind'. Improving the identity and public perception of solid waste management and its workforce is tantamount to improving the overall well-being of this sector.

Global Communities' training programmes, the formation of Hasirudala, and exposure visits are some effective examples that go to show how the livelihoods and personal development of waste pickers can be improved. Significant improvements have been measured in their health and hygiene, self-empowerment, and understanding of the larger waste management system and their value within it. This, in turn, improves their abilities to direct their futures, in their personal lives and occupations.

A Roadmap for Integrating Informal Waste Collectors in City Systems

The following steps were recommended to the BBMP through the affidavit referenced earlier to integrate waste pickers into the waste management systems of Bengaluru:

1. Register and provide photo identity cards to ragpickers, waste pickers, and itinerant waste buyers.
2. Provide financial, human, and infrastructural support to register all waste pickers.
3. Provide support to registered organizations of waste pickers for organizing and training.
4. Provide spaces in every neighbourhood for undertaking composting, biogas, and segregation.
5. Formulate welfare measures for waste pickers and provide for them in the municipal budget.
6. Promote and provide capital and infrastructure costs from the municipal budget for micro-waste collection and processing enterprises.
7. Issue trade licences where appropriate.
8. All registered waste pickers should be eligible for benefits under government schemes, irrespective of below poverty line (BPL) status.
9. Ensure that organizations of waste pickers are represented on the District Solid Waste Management Monitoring Committee.

10. Ensure that all mixed waste is first segregated for recovery of recyclables.
11. Ensure that all waste generators pay the prescribed service fees.
12. Encourage the presence of senior police officers and ensure that police understand the legal status of the work undertaken by waste pickers.
13. Regularly monitor and quantify the reduction in greenhouse gases as a result of resource recovery.
14. Include waste pickers in the collection centres for plastic proposed in the Plastic Waste Rules, 2011.
15. Include representatives of waste pickers in any discussions about them and decisions made by BBMP on their behalf and/or with respect to solid waste management.

***

The complexity of urbanization and poverty requires a fundamental rethinking of some basic assumptions about poverty and the solutions that have been tried to date. New relationships between the private sector, government, and civil society need to be forged so that each can bring to bear its particular strengths in engaging and delivering creative new solutions.

The private sector cannot do it alone, nor will markets alone solve all problems. However, the case studies of this chapter illustrate how multiple stakeholders can design solutions that harness market forces and the energy of entrepreneurship to address some of the problems facing an urbanizing India. Ultimately, the success of any solution needs to be measured by its ability to sustain, scale, and empower the urban poor by helping them gain new skills and connection with the wider economy, so that India can unlock its full economic potential.

References

Akerlof, George A. and George E. Kranton. 2010. *Identity Economics: How Our Identities Shape Our Work, Wages, and Well-Being.* New Jersey: Princeton Univeristy Press.

Chakraberty, Sumit. 2014. 'How Bangalore's New Recycling Plan Helps Its Poorest Residents', *The Atlantic*, 15 April. Available at http://www.citylab. com/work/2014/04/how-bangalores-new-recycling-plan-helps-its-poorest-residents/8887/, last accessed on 20 April 2017.

CHF International and Mythri Sarva Seva Samithi. 2010. 'Sample Study of Informal Waste Pickers in Bangalore: September–October 2010'. Available at https://www.globalcommunities.org/publications/Sample%20Study%20 of%20Informal%20Waste%20Pickers%20in%20Bangalore.pdf, last accessed on 20 April 2017.

Gupta R., S. Sankhe, R. Dobbs, J. Woetzel, A. Madgavkar, and A. Hasyangar. 2014. *From Poverty to Empowerment: India's Imperative for Jobs, Growth, and Effective Basic Services*. Bangalore: McKinsey Global Institute.

Kapur, A. 2011. 'Indian Scavengers doing what Officials can't', *The New York Times*, 19 January. Available at http://www.nytimes.com/2011/01/20/ world/asia/20iht-letter20.html?_r=0, last accessed on 20 April 2017.

Karnataka State Pollution Control Board (KSPCB). 2013. *Municipal Solid Waste Annual Report*. Available at http://kspcb.kar.nic.in/MSW%20Annual%20 Report%20%202013-14%20.pdf, last accessed on 20 April 2017.

Ministry of Housing and Urban Poverty Alleviation (MHUPA). 2013. *State of Slums in India: A Statistical Compendium*. Available at http://www.nbo.nic. in/Images/PDF/Slums_in_India_Compendium_English_Version.pdf, last accessed on 8 June 2017.

Ministry of Urban Development (MoUD). 2000. *Manual on Solid Waste*. Available at http://moud.gov.in/swm_manual, last accessed on 20 April 2017.

———. 2016. *Municipal Solid Waste Management Manual*. Central Public Health and Environmental Engineering Organisation (CPHEEO). Available at http://moud.gov.in/pdf/57f1e55834489Book03.pdf, last accessed on 20 April 2017.

National Skill Development Corporation (NSDC). 2013. *Human Resource and Skill Requirements in the Building Construction and Real Estate Sector*. Available at http://www.nsdcindia.org/sites/default/files/files/Building-Construction-Real-Estate.pdf, last accessed on 20 April 2017.

Prahalad, C.K. and Harvey Fruehauf. 2004. The Fortune at the Bottom of the Pyramid. Wharton School Publishing.

Sankhe, S., I. Vittal, R. Dobbs, A. Mohan, A. Gulati, J. Ablett, S. Gupta et al. 2010. *India's Urban Awakening: Building Inclusive Cities, Sustaining Economic Growth*. Bangalore: McKinsey Global Institute.

Sharma, R. and S. Chitkara. 2006. 'The Informal Sector in the Indian System of National Accounts', Expert Group on Informal Sector Statistics, Paper No. 6, Ministry of Statistics and Programme Implementation, India.

The Times of India. 2014. 'Shut Down Mandur Landfill, Pollution Control Board Says', 1 June. Available at http://timesofindia.indiatimes.com/city/bengaluru/Shut-down-Mandur-landfill-pollution-control-board-says/articleshow/35868463.cms, last accessed on 20 April 2017.

Conclusion

This volume emerges from the profound and diverse ideas on innovation in India explored at the University of Pennsylvania's conference, India as a Pioneer of Innovation: Constraints and Opportunities, held in November 2013. The rich tapestry and wide canvas of innovative practices in India constructed by panellists at the conference have, largely, been reflected in this volume. India's unique background conditions—democratic governance, massive human population, emerging middle class, weak government, huge informal economy—and the diversity of innovative practices that work around existing structural, behavioural, and systemic barriers to entrepreneurship lend tremendous novelty to the narratives on innovation presented here.

Of the many insights offered, if there is one that perhaps defines this volume, it is the context-specificity of innovation. Scholars often ignore this truth in their attempt to build generalized theories and objectively measurable indices of innovation. The direction taken by innovation is highly dependent on various factors such as temporal and spatial context, regulatory framework surrounding the same, nature of public demand that drives solutions, and resources available at hand. There cannot possibly be any one model or index of innovation to accurately measure a country's innovative potential. Whether it be innovative practices followed by merchants in the colonial past, or the wide variety of innovative solutions at play in India's informal sectors, terms such as 'Indovation', *jugaad*, and frugal innovation are, at best, fractional successes in capturing the complexity of innovation here.

But being aware of this, it is for India to re-brand herself as an innovator country, and highlight the manifold ways in which solutions are offered to end users. The anecdotes may vary, ranging from innovative activity in healthcare (Chapter 3), family business groups (Chapter 4), 'innovations for the millions' (Chapter 8), or poverty management solutions (Chapter 9), but the emphasis must consistently be on the common themes running through all of them: (a) creative ways of addressing problems, and (b) India's future as being closely tied with this 'attitude' of innovation. It is important, therefore, to highlight India's innovative past as done here, as well as project globally some of the recent initiatives, both private and public.

Towards creating digital rails for use by start-ups, the Government of India (GoI) has worked in partnership with private entities to develop IndiaStack—a set of open Application Programming Interfaces (APIs). The stack, which relies on India's most comprehensive citizen identification database, Aadhaar, for verification and authentication, allows digital start-ups to plug in and use the various services—payment interface, digital lockers, consent-based access to private details—on offer instead of developing them all by themselves. New start-ups are addressing India's poor financial inclusion record through innovative ideas including data-based lending and near-field communication devices for payment in locations without Internet access. Some of India's recent policy innovations too are worth noting. The idea of payment banks, originally mooted by the Reserve Bank of India, has now gained traction with the award of licences to digital wallet players, telecommunication service providers, and India Post (the Department of Posts). The ongoing Indo-Japanese collaboration should soon result in successful completion of the promised Mumbai–Ahmedabad high-speed bullet train project. Hyperloop technology, designed to propel a pod-like vehicle through a reduced-pressure tube, is awaiting green signal from GoI. The Ministry of Power is doing a commendable job in promoting renewable energy sources for electrification, including solar and wind farms.

However, while we do laud private and public entities for overcoming many of the bottlenecks highlighted in this volume, we must never lose sight of how different things could have been if such bottlenecks were neutralized in the first couple of decades after Independence. Institutional voids, whether in the form of difficulties or, often total absence, of last mile reach; vagaries of water, power, and Internet supply; inadequate machinery for contract enforcement; or robust credit access at affordable rates for small and medium-sized entrepreneurs, strangle innovative potential. Presidential proclamation of an entire decade (2010 to 2020) as India's decade of innovation, releasing policy documents advocating an integrated approach to science, technology, and innovation, or even creating new government departments and think tanks evangelizing technological disruption, do not deliver tangible outcomes despite their signalling of progressive intentions. Many of the aforementioned voids continue to persist, and need to be addressed on war footing.

The confusing, and often inconsistent, regulatory and policy stances are also a huge impediment to innovation. Consider, for example, the Indian regulator's approach to the use of civilian drones. From 2014, there has been a blanket ban on flying drones, resulting in a choking of funds to young start-ups in this evolving area of technological advancement. Instead, the Directorate General of Civil Aviation could have learnt from some other regulators, such as those in the UK, who are proactively working with experimental drone use for delivery and other cases. Similarly, a nation that proudly endorsed its financial technology start-ups including digital wallet companies post the demonetization of certain currency denominations in November 2016, started sending out mixed regulatory signals regarding these innovations within less than six months from the date of this decision. The Reserve Bank of India's draft master directions dated 20 March 2017 propose a host of onerous requirements—minimum positive net worth of INR 25 crore; full compliance with electronic know your customer (KYC) requirements; maximum monthly cap of INR 10,000 on fund transfers from mobile wallets—which disproportionately prioritize safety over innovation without appreciating the business model.

It would therefore bode well for India to stand reminded that the country places 130th in the World Bank Group's 2016 Ease of Doing Business rankings. This reality also explains the lopsided spurt in online portals and digital start-ups as compared to ventures in the manufacturing space. The former benefit from the minimal presence of both ex ante licences prior to commencement of business, and regulations governing their day-to-day activity. But growing the economy in a more sustainable fashion requires attracting talent and ideas to other sectors too, and regulatory disparities serve no one involved in this mission.

India also needs to evolve in terms of quicker and more efficient turnaround of policies to actual outcomes. This is particularly so with innovation because the sluggish pace can mar entire industries and technologies. With the laudable objective of promoting start-up activity, GoI announced a government fund of INR 100 bn for financing tech startups, in July 2014. Nearly three years have lapsed with minimal fund disbursement. The primary reason: unrealistic qualifying parameters imposed on beneficiaries and joint investors under this scheme. Similarly, several policy recommendations have already been put forth to address well-chronicled barriers to innovation in the Indian biopharma and bio-agriculture sectors, including the need for single-window clearances for biodiversity exploration, preliminary testing, clinical/field trials, and final approvals. Yet, over a period of more than 15 years starting from 2001, GoI has been unable to put such mechanisms in place.

It is important to realize, and quickly, that innovation is often an exercise in risk-taking, and the nation, as much as innovators, must be willing to take them rather than design static policies that appear foolproof, but fail to deliver results. Most critically, such progressive mentality must go beyond thinking and action at the central government level to the various state governments and their agencies. For instance, in the online cab aggregator industry, a committee set up by the Ministry of Road Transport, GoI, has recommended cutting the regulatory cord with progressive free-market proposals. Despite this committee's report that came out in December 2016, some of the state governments continue to throttle industry players using their authority under the legacy enactment, the Motor Vehicles Act, 1988. A better system of coordination and information exchange between the centre and the states lies at the core of developing progressive policies.

Innovation in today's world also entails growing new markets through nation-wide incubation of emerging technologies. The increasing and open competition between nation states to project themselves as hubs of early innovation in cutting-edge areas of technology—artificial intelligence, synthetic biology, quantum computing, etc.—is a trend India cannot ignore. To an extent, the nation has already realized this, as evidenced from the National Intellectual Property Rights Policy, 2016 and the grand challenges, hackathons and contests now regularly conducted by both the Centre and States to propel new ideas and solve big problems. But growth in framework technologies needs an altogether different approach. India does not yet have an equivalent of a body like the US Defense Advanced Research Projects Agency (DARPA) which has been responsible for incubating several path-breaking innovations, including the Internet, advanced robotics, and driverless vehicles. As argued in Chapter 5, the biopharma sector too needs an aggressive policy push rather than mere regulatory responses. More so when potential game changers including CRISPR[1] are being developed in universities and private research institutions abroad.

World Bank data, last available for 2013, places India's research and development (R&D) spending as somewhere between 0.8 per cent and 0.9 per cent of its GDP, much lower than countries, such as China (2 per cent), Japan (3.5 per cent), South Korea (4.15 per cent), and Israel (4.1 per cent). The industry and private sector in these countries contribute 60 per cent to 70 per cent of this research expenditure. Indian industry's share falls way behind, fluctuating around 30per cent. Thus, both government and private actors are lagging in their commitment to R&D. In addition, a major challenge for the nation is to foster a culture of innovation. For a society dominated by conventional thinking, rigid social structures, and conformist expectations from its youth, making this change requires a total revamp of the education system at all levels.

This volume's silence on the theme of education should not in any way be construed as undermining its significance to the larger innovation project. Some of the recent initiatives launched by the Government of India are worthy of mention. The Atal Innovation Mission (AIM), along with the Self Employment and Talent Utilization (SETU) scheme administered by NITI Aayog, are meant to facilitate school-level financial grants. The AIM has started promoting Atal Tinkering Laboratories (ATLs) in schools across India, to inculcate a design mindset, computational thinking, adaptive learning, physical computing, and other skills

---

[1] Clustered regularly interspaced palindromic repeats, a new technique of gene editing.

vital to an innovation culture. However, excessive regulations pertaining to education at all levels in India remain a problem. These regulations prioritize deliverables over actual learning outcomes and restrict institutional flexibility in designing creative teaching methods.

To illustrate with an example, data science can create several good jobs in the coming years and perhaps even compensate for the automation-induced job losses in India's information technology services sector. But the nation is severely ill prepared to take advantage of this opportunity. Training in data science demands agile curriculum-setting and higher levels of industry–academia engagement, both missing in several educational institutions within the country. In addition, access to public data is often a big constraint, thanks to the way the right to information law is structured. These are issues best addressed through changes in policy thinking rather than market-based solutions. If data is the new oil, we certainly need more of it to fuel the innovation engine.

To conclude, there are clear consequentialist undertones to any kind of innovation in India, because the country and its unique context make it an avid and persistent searcher for results. Governments, policymakers, and businesses are largely confined within this box of consequences and end up innovating within its boundaries under the present scenario. This is not to say that the country should, or can afford to, ignore tweaking with existing structures and designs for purposes beyond functionalism and results. To build a sustainable society in the long run, innovation bereft of any immediately foreseeable consequences must be strongly encouraged. Innovation of this kind, done for its own sake, facilitates a more organic transition to a creative society with indigenous solutions. The state and private actors must jointly work towards this transition. But even when doing so, context cannot be ignored, as demonstrated by the chapters here. The present Modi government at the centre has initiated the 'Make in India' campaign to give a boost to manufacturing and technological innovation in the country. Such laudable efforts can only fulfil their projected outcomes if they are acutely sensitive to the localized context surrounding the Indian innovation ecosystem at present.

Index

About the Editors and Contributors

Editors

Harbir Singh is the Mack Professor of Management, the co-director of the Mack Center for Technological Innovation, and the vice dean of global initiatives at The Wharton School, Pennsylvania, USA. His work focuses on strategic alliances, corporate acquisitions, and corporate restructuring and he serves on the editorial boards of the *Strategic Management Journal* and the *Academy of Management Journal*. Singh is also a fellow of the Strategic Management Society, Chicago, USA, and is a former division chairperson of the Business Strategy Division of the Academy of Management, New York, USA. He is on the Academic Advisory Board of the Indian School of Business, Telangana, India, and has been a visiting professor at the London Business School, UK, and Bocconi University, Italy.

Ananth Padmanabhan is a technology and policy fellow at Carnegie India, New Delhi, and a doctoral student enrolled on non-resident status at the University of Pennsylvania Law School, USA. He works at the intersection of technology, regulation, and public policy, and is the author of the leading treatise, *Intellectual Property Rights: Infringement and Remedies* (2012). He practiced in the High Court of Madras for more than five years before transitioning to a career in academia. He graduated from the National Law School of India University, Bengaluru, India (2007), holds a master's degree in law from the University of Pennsylvania Law School, USA (2014), is winner of the Karin Iest Award for the best outgoing student in the LLM batch, and is currently enrolled in their doctoral programme on a non-residence basis.

Ezekiel J. Emanuel is the vice provost for global initiatives; the Diane v.S. Levy and Robert M. Levy University Professor; and the chairperson of the Department of Medical Ethics and Health Policy at the University of Pennsylvania, USA. He was the founding chairperson of the Department of Bioethics at the National Institutes of Health, Maryland, USA until August of 2011. He recently served as a special advisor on health policy to the director of the Office of Management and Budget, USA, and National Economic Council, USA. Emanuel has been published widely on the topics of healthcare reform, research ethics, and end-of-life care. His recent books include *Reinventing American Health Care: How the Affordable Care Act will Improve our Terribly Complex, Blatantly Unjust, Outrageously Expensive, Grossly Inefficient, Error Prone System* (2014) and *Prescription for the Future: The Twelve Transformational Practices of Highly Effective Medical Organizations* (2017).

Contributors

Shyamkrishna Balganesh is a professor of law, a co-director at the Center for Technology, Innovation and Competition, and a co-director of the Center for Asian Law, at the University of Pennsylvania Law School, USA. His research focuses on understanding how intellectual property law and innovation policy can benefit from the use of ideas, concepts, and structures from different areas of the common law, especially private law. Some of his recent work attempts to understand the working of India's intellectual property laws and the ways in which they can be integrated fruitfully with India's economic and cultural development goals.

Chirantan Chatterjee is a faculty member in economics and public policy at the Indian School of Business, Telangana, India, where he is also the Bharti and Max Research Fellow in Public Policy and Healthcare. His research interests are in empirical industrial organization, economics of innovation, and pharmaceutical economics. Chatterjee's work has been published in top peer-reviewed journals including the *RAND Journal of Economics, Journal of Health Economics, The Journal of Law & Economics, Research Policy,* and *Social Science & Medicine*. In the past, Chatterjee's research has been supported by the National Science Foundation, Pfizer, and Qualcomm. He has also consulted with the World Bank and the Competition Commission of India, and his inputs have been cited in a 2015 Tarun Khanna Committee Report by NITI Aayog on innovation and entrepreneurship policy for the Indian economy.

Brian English is an urban planner who works at the intersection of sustainability, technology, and economics. He is the director of the Global Cities Initiative at Global Communities, an international development organization, and is also a fellow at Institute for Urban Research, University of Pennsylvania, USA. Previously, he was the country director for Global Communities in India and directed a USD 6 million programme called SCALE-UP funded by the Bill and Melinda Gates Foundation to reduce urban poverty. After Hurricane Ivan in 2004, English managed community revitalization programs in the Eastern Caribbean for the United States Agency for International Development.

Budhaditya Gupta is a lecturer of management at the University of Melbourne, Australia. His research explores how organizations design, develop, and manage their resources and processes as they pursue innovation and entrepreneurship initiatives. As part of this research agenda, Gupta has worked with Tarun Khanna at Harvard University, Massachusetts, USA, to study (a) the task-shifting-based cardiac surgery staffing model at Narayana Health (NH) in India, published in *Healthcare: The Journal of Delivery Science and Innovation* and (b) the development of a tertiary care hospital in the Cayman Islands by NH to serve patients from the USA and Caribbean Islands, a case study published by *Harvard Business Publishing*.

Barbara Harriss-White is Emeritus Professor of Development Studies, Emeritus Fellow of Wolfson College, University of Oxford, UK, and a visiting professor at Jawaharlal Nehru University, New Delhi, India. Since 1969, her field research in India on rural development and aspects of deprivation has, with others, generated over 250 papers and chapters and 35 books. In 2009, her book, *Rural Commercial Capital: Agricultural Markets in West Bengal*, won the Edgar Graham Prize.

Prashant Kale is Associate Professor of Strategic Management at the Jesse H. Jones School of Business, Rice University, Texas, USA. His research focuses on strategic alliances, mergers and acquisitions, and emerging economies and he has been published in reputed international journals such as the *Strategic Management Journal, Harvard Business Review, California Management Review, MIT Sloan Management Review, European Management Journal, Managerial and Decision Economics,* and *Academy of Management Perspective*. He is a member of the editorial boards of the *Strategic Management Journal, Journal of International Business,* and *Strategic Organization*. He has received several awards for excellence in teaching and he was also rated among the Top 10 Business School Professors by *Bloomberg Businessweek*.

Tarun Khanna is the Jorge Paulo Lemann Professor at the Harvard Business School, and the first director of South Asia Institute at Harvard University, Massachusetts, USA. He has taught courses related to entrepreneurship and economic development at Harvard for over two decades. He is a young global leader of the World Economic Forum, a fellow of the Academy of International Business, Michigan, USA, and was recently recognized by the Academy of Management, New York, USA, for lifetime scholarly achievement. Outside of Harvard, he is a co-founder of several entrepreneurial ventures across emerging markets, and serves on several boards of directors. In 2015, he was chosen by the Government of India to chair a national committee to help shape the fabric of India's entrepreneurial ecosystem.

Vijay Mahajan is the founder and CEO of the Basix Social Enterprise Group, Hyderabad, India, which has, since 1996, supported the livelihoods of over three million poor households. In 1983, he founded PRADAN, an NGO engaged in livelihood promotion. He serves on the boards of a number of NGOs and on several major advisory committees constituted by the Government of India. He was named a distinguished alumnus of the Indian Institute of Technology Delhi and the Indian Institute of Management Ahmedabad. He was selected as one of the 60 outstanding social entrepreneurs of the world at the World Economic Forum, Davos, Switzerland in 2003. He mentors the Innovators' Collaborative in Hyderabad.

Shreekanth Mahendiran is a research advisor at Centre for Budget and Policy Studies, Bengaluru, India. His research primarily focuses on the applied microeconomics and development economics, where he employs insights from industrial organization, microeconomic theory, neo-institutionalism, and development literature to examine the questions on access, welfare, and impact assessment. He has been working in areas related to pharmaceutical industry, healthcare market, education, and empowerment in the context of the developing world.

Claude Markovits is Senior Research Fellow Emeritus at the Centre of Indian and South Asian Studies, École des Hautes Études en Sciences Sociales, Paris, France. He is a historian of colonial India, with a special interest in the merchant world. His publications include *Indian Business and Nationalist Politics 1931–1939* (1985); *The Global World of Indian Merchants c. 1750–1947: Traders of Sind from Bukhara to Panama* (2000), which received the A.K. Coomaraswamy Book Prize of the American Association of Asian Studies in 2002; *Merchants, Traders, Entrepreneurs: Indian Business in the Colonial Era* (2008).

David Nimmer is of counsel to Irell & Manella LLP in Los Angeles, California, USA, and a Professor of Practice at UCLA School of Law, California, USA. Since 1985, he has authored and updated *Nimmer on Copyright*, the standard reference treatise in the field, first published in 1963 by his late father, Melville B. Nimmer. He has authored numerous articles about American and international copyright law and policy. He also gave congressional testimony at the invitation of the House Judiciary Committee in 2014; on behalf of the United States Telephone Association in 1997; and on behalf of the National Association of Broadcasters in 1992. He also delivered parliamentary testimony on behalf of the Combined Newspaper and Magazine Copyright Committee of Australia in Sydney in 1993.

T. Sundararaman is professor and dean of the School of Health Systems Studies at Tata Institute of Social Sciences (TISS), Mumbai, India. Earlier he served as a member of the faculty of internal medicine in Jawaharlal Institute of Postgraduate Medical Education and Research, Puducherry, India. He also served as director, state health resource center in Chhattisgarh, India and executive director of National Health Systems Resource Center, New Delhi, India (which was established under his leadership). In 2014 he returned to academics, first as visiting professor in Jawaharlal Nehru University, New Delhi, and then at TISS since April 2015.

Rajani Ved is advisor at the National Health Systems Resource Center, New Delhi, India, providing technical support to the Government of India's National Health Mission. She has worked for over 25 years in programme implementation, evaluation, research, and policy formulation in the area of health systems strengthening and women and children's health and nutrition. Her special interest is the study and documentation of scaling up innovations from small-scale NGOs and government-led pilots to state and national public health systems. She has developed tools for scaling up and conducts seminars and workshops for training on scaling up.